Sustainable Sewage Sludge Management and Resource Efficiency

Edited by Başak Kiliç Taşeli

Published in London, United Kingdom

IntechOpen

Supporting open minds since 2005

Sustainable Sewage Sludge Management and Resource Efficiency
http://dx.doi.org/10.5772/intechopen.80201
Edited by Başak Kiliç Taşeli

Contributors
B.P. Naveen, P.V. Sivapullaiah, Sridhar Pilli, Ashutosh Kumar Pandey, Ankur Katiyar, Kritika Pandey, Rajeshwar Dayal Tyagi, Marko Likon, Marjan Zemljič, Rodrigo De Freitas Bueno, Başak Kiliç Taşeli, Jarosław Górecki, Wesam Salah Alaloul, Muhammad Ali Musarat

First published in London, United Kingdom, 2020 by IntechOpen
IntechOpen is the global imprint of INTECHOPEN LIMITED, registered in England and Wales, registration number: 11086078, 7th floor, 10 Lower Thames Street, London, EC3R 6AF, United Kingdom
Printed in Croatia

British Library Cataloguing-in-Publication Data
A catalogue record for this book is available from the British Library

Additional hard and PDF copies can be obtained from orders@intechopen.com

Sustainable Sewage Sludge Management and Resource Efficiency
Edited by Başak Kiliç Taşeli
p. cm.
Print ISBN 978-1-83962-706-4
Online ISBN 978-1-83962-707-1
eBook (PDF) ISBN 978-1-83962-708-8

Meet the editor

Dr. Başak Kiliç Taşeli is a professor in the Environmental Engineering Department at Giresun University, Turkey. She is an expert with 27 years of experience in environmental management, waste prevention, renewable energies, and zero waste. She is an environmental engineer and completed her MSc in soil pollution due to thermal power plants and her PhD in environmental sciences-biodegradation of toxic organic compounds at Middle East Technical University. She teaches graduate-level sustainability courses concerning environment, energy and sustainability; biofuels; biomass conversion systems; biogas production and usage; and sustainable operation of treatment plants. She also teaches undergraduate-level courses covering topics such as water supply; wastewater disposal; environmental and engineering ethics; environmental ecology; sludge treatment; unit operations; and industrial wastewater treatment. She has more than 25 years of experience working in national/international environmental projects on water quality monitoring and conservation of specially protected areas, design and operation of wastewater treatment plants, environmental impact assessment, management plans, carrying capacity, and Turkish Grand National Assembly Research and Investigation Commission Reports on various environmental pollution issues. Dr. Taşeli has excellent knowledge of zero waste, ecological sanitation, and greenhouse gas emissions calculations. Additionally, she has strong knowledge of climate change–greenhouse gas emissions relationships, horizontal subsurface flow constructed wetland's greenhouse gas emission calculations, and wastewater treatment plant's sustainability and their effect on climate change, as well as strong knowledge of international environmental standards (ISO 14001, EMAS). Finally, she has 10 years of experience working in EU-funded Erasmus + projects as promoter and partner.

Contents

Preface **XI**

Section 1
Sustainable Sewage Sludge Management **1**

Chapter 1 **3**
Does Sustainable Management of Biodegradable Sludge Exist at All?
A BACOM Project Case
by Marko Likon and Marjan Zemljič

Chapter 2 **23**
Pre-treatment Technologies to Enhance Anaerobic Digestion
by Sridhar Pilli, Ashutosh Kumar Pandey, Ankur Katiyar, Kritika Pandey and Rajeshwar Dayal Tyagi

Chapter 3 **51**
Composition, Production, and Treatment of Sewage Sludge
by Rodrigo de Freitas Bueno

Chapter 4 **65**
Solid Waste Management: Current Scenario and Challenges in Bengaluru
by B.P. Naveen and P.V. Sivapullaiah

Chapter 5 **89**
Comparative Life Cycle Assessment of Sewage Sludge (Biosolid) Management Options
by Başak Kiliç Taşeli

Section 2
Green Economy **99**

Chapter 6 **101**
Circular Economy and Green Public Procurement in the European Union
by Jarosław Górecki

Chapter 7 **123**
Impact of Zero Energy Building: Sustainability Perspective
by Wesam Salah Alaloul and Muhammad Ali Musarat

Preface

During the last 50 years, the world has witnessed strong economic growth and fast urbanization rates, which have lifted millions of people out of poverty. However, this development has been coupled with increasing use of resources and environmental degradation. It is obvious that the increased use of resources is strongly coupled to environmental impacts. Partly this is because of the linear way resources are used. The process leads to environmental impacts all the way and ends with different kinds of waste. Creating decent living conditions for all people while decoupling economic growth from the increasing use of virgin resources and environmental impacts is the major challenge of this millennium. This is also the essence of the United Nations' 2030 Sustainable Development Goals. There are many approaches suggested for solving these problems. One is to change consumption behavior from material products to services. Another option is to find technological solutions to create more closed loops for materials and use fewer virgin resources and energy obtained from clean renewable sources.

Wastewater treatment plants are aimed at decreasing the environmental impacts of discharging untreated wastewater into receiving bodies. Considering the need for long-term ecological sustainability, the objectives of wastewater treatment systems need to include energy and resource savings and waste reduction. Sewage sludge management is a management system that makes sludge recovery a central component of a wastewater treatment plant that strives to integrate it with improving overall sustainability of the plant.

Sludge formation during wastewater treatment is inevitable even with proper management and treatment. However, proper treatment and disposal of sludge is still difficult in terms of cost, the presence of new pollutants, health problems, and public acceptance. Conventional disposal methods (e.g., storage, incineration) have raised concerns about legislative constraints and community perception that encourage the assessment of substitute sludge-management options. Sludge management requires a systematic solution combining environmental effectiveness, social acceptability, and economic affordability. Increased production of sludge (biosolids) worldwide is due to population growth, urban planning, and industrial developments. The sludge needs to be properly treated and environmentally managed to reduce the negative effects of its application or disposal.

The first aim of this book is to investigate the application of biosolids or sewage sludge, together with possible resources for sustainable development. The second aim is to view resource efficiency from a more complex perspective looking at several resources and the causal links between them in order to point out new pathways towards a more sustainable use of resources. The book consists seven chapters divided into two sections: "Sustainable Sewage Sludge Management" and "Green Economy."

The editor would like to acknowledge and thank the authors for their contributions.

Dr. Başak Kiliç Taşeli
Professor,
Department of Environmental Engineering,
Giresun University,
Turkey

Section 1

Sustainable Sewage Sludge Management

Chapter 1

Does Sustainable Management of Biodegradable Sludge Exist at All? A BACOM Project Case

Marko Likon and Marjan Zemljič

Abstract

Due to the modern lifestyle and the formation of large amounts of biodegradable sludge, its processing is becoming a demanding technological and logistical project. Stabilization with pozzolanic ash and its reuse in construction industry represents one of the possible sustainable solutions. Mixing biodegradable sludge with pozzolanic ash triggers a set of physiochemical reactions such as converting heavy metals into insoluble hydroxides, forming heat due to hydration of metal oxides, and forming of a set of pozzolanic structures due to high pH and heat. Studies showed that the produced material is biologically and chemically inert and safe for use under controlled conditions. Comparison of different most widely used technologies, using life cycle analysis, indicated advantages of using material conversion of biodegradable sludge into materials rather than using it for energetic purposes. Based on the calculation of their negative influence on the environment and human health, the analyzed technologies can be categorized from those with less impact to those with higher impact: stabilization with ash < pyrolysis < anaerobic digestion < composting < landfilling. The life cycle assessment (LCA) showed that the decentralized technologies enabling material use of biodegradable sludge are more sustainable than centralized installations for composting biodegradable sludge in large quantities.

Keywords: biodegradable sludge, sustainable, stabilization, pozzolanic ash, LCA

1. Introduction

The management of biodegradable sludge (BS) is becoming an important challenge for developers, investors, and managers of wastewater treatment plants (WWTPs) over the world. BS is an unavoidable by-product of advanced techniques of biological purification of wastewater. In general, global population generates between 70 and 90 g BS/person or 1 ton of dry substance/10,000 persons [1]. The management of BS is not only a trivial problem, and investors should be aware about it during the designing of the WWTPs, because studies made in Austria showed that the management of BS (stabilization, drying, pretreatment, and transport) can easily exceed 40 to 53% of all operational costs during the processing of waste waters [2]. Proper and sustainable management with biodegradable waste is the outmost important topic in developed countries, because more expert studies showed that inappropriate approach to management with those waste includes high risks for human health and environment. Legislators all over the world are promoting methods and techniques allowing decreasing amount of BS on sanitary landfills with the aim of decreasing the greenhouse emission as well as decreasing the quantity of landfill leachate [3–6]. In accordance with Council Directive on the landfill of waste (1999/31/EC), landfilling of BS on sanitary landfills is forbidden within the EU-28 countries from July 2009. The Directive on waste from 2008 (2008/98/EC) even more restrictively implements hierarchy with which the European Commission is trying to promote material and energetical use of waste.

Nevertheless, blindly following the Directives does not necessarily result in the implementation of optimal and sustainable solutions, and we need to think about new approaches. However, because of those new approaches, all decisions need to be supported by different scientific data, and after that all individual solutions need to be evaluated on the same basis.

In order to allow equal treatment and consistent evaluation of different approaches in the management of BS, the EU Joint Research Centre has developed the "life cycle thinking" (LCT) and "life cycle assessment" (LCA) methodologies and has given the interpretation/instructions on how to use these methods, which should facilitate the decision-making and help managers to establish the most effective and sustainable management system for BS treatment [7]. The guidelines were prepared in cooperation with the International Organization for Standardization (ISO) and are registered under the number SIST EN 14040 and 14,044 [standards for LCA and the International Reference Life Cycle Data System (ILCD) Handbook]. A similar approach in the assessment is used in most European countries and almost by all authors of professional publications.

Each year, EU-28 member states produce more than 10 Mt. of communal BS calculated on dry matter [8, 9]. According to the official data, more than 40% of BS is still being disposed on municipal waste landfills [10]; approximately 36% are reused in agriculture or incinerate [8]; however, Eurostat has no data on the remaining 24% BS [7]. Biodegradable sludge is generated during the operation of urban WWTP utility and contains heavy metals, poorly dissoluble organic compounds (residues of detergents, washing agents, personal hygiene preparations, medicines, etc.), and possibly pathogens microorganisms (**Tables 1–4**).

Both unprocessed and processed BS contains heavy metals, residues of phytopharmaceutical products, and surfactants. Their content depends on the origin of the municipal sludge (sludge from the municipal, industrial, or combined treatment plant). In addition to active microorganisms, pathogenic bacteria listed in **Table 3** are also present in unhygienic feces. Because of that, it is difficult to compost them, and the manufactured compost is prohibited for uses in agriculture as fertilizer.

The method for the management of BS is different and depends on their origin, composition, especially contents of hazardous and biodegradable substances, and on available infrastructure and local regulations. BS, which is disposed on landfills, is subject to uncontrolled aerobic and anaerobic processes that cause the release of a large amount of greenhouse and noxious gasses (CH_4, CO, CO_2, H_2S, etc.) and emissions of heavy metals. Currently, the EU and the USA are using incineration, composting, stabilization, and landfilling as recognized methods for bio sludge disposal. All mentioned methods have negative environmental impacts, especially landfilling and incineration. The special problem for the environment presents pathogenic microorganisms of different species that exist inside biodegradable sludge and which must be stabilized or neutralized before further application. One of the most economical and environmentally accepted methods for stabilization of BS is their mixing with wasted alkaline materials as ash, slag, foundry sand, and foundry dust.

According to the available data from the literature and the LCA analyses, the disposal of BS is the worst choice between the possible solutions, even if landfills are equipped with gas capturing systems and devices for its energetic use [12–14].

Constituent	Unit	Unprocessed BS		Processed BS		Active BS
		Range	Avg.	Range	Avg.	
Dry solid	% d.s.	2.0–8.0	5.0	6.0–12.0	10.0	0.83–1.16
Volatile sub.	% on d.s.	60–80	65	30–60	40	59–88
Fats and oils	% on d.s.	6–30	—	5–20	18	5–12
Proteins	% on d.s.	20–30	25	15–20	18	32–41
Nitrogen (tot)	% on d.s.	1.5–4.0	2.5	1.6–6.0	3.0	2.4–5.0
Phosphorous	% on d.s.	0.8–15.0	1.6	1.5–4.0	2.5	2.8–11.0
Ash (K_2O)	% on d.s.	0–1	0.4	0.0–3.0	1.0	0.5–0.7
Cellulose	% on d.s.	8.0–15.0	10.0	8.0–15.0	10	—
Iron	% on d.s.	2.0–4.0	2.5	3.0–8.0	2.5	—
Silicates (SiO_2)	% on d.s.	15.0–20.0	—	10.0–20.0	—	—
Alkalis ($CaCO_3$)	% on d.s.	500–1500	600	2500–3500	—	580–1100
Organic acids	% on d.s.	200–2000	500	100–600	3000	1100–1700
Energetic value	TJ/ton	10–12.5	11	4–6	0.2	8–10
pH value		5.0–8.0	6	6.5–7.5	7	6.5–8.0

Table 1.
Average composition of BS in Europe—combined WWTPs [1].

Constituent	Range	Average	Unit
Cr	20–60	35	mg/kg of dry solid
Cu	200–600	375	mg/kg of dry solid
Pb	100–400	175	mg/kg of dry solid
Ni	15–50	30	mg/kg of dry solid
Sb	1–5	3	mg/kg of dry solid
Zn	500–1500	900	mg/kg of dry solid
As	5–20	12	mg/kg of dry solid
Hg	0.5–3	1.4	mg/kg of dry solid
Cd	1–5	2	mg/kg of dry solid
Mo	4–20	8	mg/kg of dry solid

Table 2.
Average composition of heavy metals in sewage sludge in Europe—combined WWTPs [11].

Disposal of 1 ton of BS with 20% dry substance emits up to 296.9 kg of $CO_{2(eq)}$ into the atmosphere.

The incineration of BS which is generated by the operation of WWTPs is therefore becoming a common practice of management with BS. The incineration is primarily used for the reduction of the volume and not for energy production, because ash represents only about 30% of the dry matter volume in BS [15]. However, ash disposal remains a serious problem as it still contains heavy metals. LCA analyses showed that the incineration of BS is meaningful only in cases where the systems of the so-called industrial symbiosis exist, such as the co-incineration of BS [15] with coal [16] but with a presumption that the BS is sufficiently dry and the incineration chamber is specially designed (FBR). The incineration of 1 ton of

Pathogen	Disease(s) and/or symptoms
Salmonella spp.	Salmonellosis, typhoid
Shigella spp.	Bacillary dysentery
Escherichia coli (enteropathogenic strains)	Gastroenteritis
Pseudomonas aeruginosa	Otitis externa, skin infections (opportunistic pathogen)
Yersinia enterocolitica	Acute gastroenteritis
Clostridium perfringens	Gastroenteritis (food poisoning)
Clostridium botulinum	Botulism
Bacillus anthracis	Anthrax
Listeria monocytogenes	Listeriosis
Vibrio cholera	Cholera
Mycobacterium spp.	Leprosy, tuberculosis
Leptospira spp.	Leptospirosis
Campylobacter spp.	Gastroenteritis
Staphylococcus	Impetigo, wound infections, food poisoning
Streptococcus	Sore throat, necrotizing, fasciitis, scarlet fever

Table 3.
List of pathogens found in BS originated from combined WWTPs.

Pollutant in sewage sludge	Domestic use	Combined sewage system	Industrial discharges
Pathogens	Human metabolism	Animal faces	Meat industry
Heavy metals	Paints (Pb), amalgam filling (Hg), thermometers (Hg), pipe corrosion (PB, Cu), batteries (Ni, Cd, Pb)	Rain (Pb, Cd, Zn), tires (Cu, Cd), roof corrosion (Zn, Cu), oil (Pb)	Various
Persistent organic pollutants	Paints, solvents, medicines, wood, treatment, cosmetics, detergents, etc.	Oil, pesticides, tar, road deicing, rain, combustion	Various

Table 4.
List of origin of different pollutants and pathogens in BS.

BS contributes 232.2 kg of $CO_{2(eq)}$ on climate change, pyrolysis up to 146.1 kg of $CO_{2(eq)}$, while burning of equal quantities of BS and energy-rich RDF can reduce pressure on climate change for 15.4 kg $CO_{2(eq)}$ [17].

The LCA itself is dependent on the environment as well on economic, social, and political conditions where the studied example is positioned. Hospido et al. [18] had studied and conducted a comparative study of agricultural use, incineration, and pyrolysis of BS and came to the conclusion that the ecologically acceptable solution is the co-incineration of BS with a coal, but at the same time, this option is least economically acceptable. It has been established that maximum efficiency and minimal environmental impact is achieved when 10–40% of dry BS are added to coal [15]. Different authors have shown practical examples where environmental impacts are mostly reduced when BS are used as fuel and produced ash used as a binder in cement production [19] and as a binder for roofing production [20].

Notwithstanding, at the EU level, composting is the most common way of managing with BS, although in the last few years, there has been made a big step towards the implementation of anaerobic degradation and energetic use of produced biogas [21, 22].

The material use of BS, composting, and anaerobic digestion with energetic use and disposal of compost on sanitary landfills seems to be the most acceptable solution [14], because large plants can be subject of discontent on heavily populated areas due to the emission of stench, bioaerosols, and heavy cargo traffic and the associated negative impacts on the environment.

These problems can be partially resolved with decentralization, namely the construction of smaller local but still economically acceptable systems with production capacity of less than 3000 tons per year [14, 23].

The advantages of the decentralized system are (1) shorter transport routes, which mean reduction of the transport costs, reducing emissions into the atmosphere and reducing noise and freight transport; (2) reducing the amount of storage of BS and consequently the reduction of stench emissions; (3) acquisitions for the local community (e.g., the exploitation of the heat produced in biogas incineration processes, which is impossible to transport on greater distances); and (4) use of smaller plants, which are attached to the environment and are less likely to be noticed. The usability of a decentralized approach by different authors has been confirmed with the introduction of the project named ForBiogas in Bologna [14].

The implementation of technologies for the material processing of BS into value-added products, which are also economically acceptable, is totally complied with the abovementioned guidelines.

This is an example of Eco-Bis technology, developed by a company GreenLife GmbH from Austria and which enables the production of bio-charcoal from BS. Particularly, this is the process of efficient drying of biodegradable sludge using vacuum filtration and subsequently pyrolysis of BS into bio-charcoal which is further used as a supplement to improve the quality of the soil. Bio-charcoal acts as a retainer for fertilizers, pesticides, and water and, at the same time, acts as fertilizer [24]. This standard is also close to BACOM technology (Biosludge Alkaline Composite Material), developed by the Slovenian company Insol d.o.o. [25].

The technology is based on the stabilization of BS, trough mixing pozzolanic ash, or other waste with alkaline properties. The final product of BACOM technology is a water-impermeable material, which can be used for as a substitute for clay in the construction and closure of different types of landfills, the construction of the beds for sewage systems and roads, the construction of inner filler of anti-flooding barriers and restoring of degraded landscapes, and the closure of degraded areas [25]. This technology is also suitable for natural storage of phosphorous. Additional drying is not necessary because ash reacts with moisture present in BS. Hydration of active metal oxides present in ash enables their conversion into alkalis. Due to the high pH and rising temperature, the heavy metals convert into water-insoluble hydroxides and chelate what prevent their further extraction. Released heat and high pH destroy pathogenic and other microorganisms which make the mixture biologically stable.

As we can see, there are different approaches and ideas about managing BS. As part of our study, we chose the typical technologies that are currently being operated and compared them with decentralized technologies that enable the material use of BS.

1.1 BACOM technology: mechanism

The BACOM technology base on the stabilization of BS, where the BS is mixed with ash and converted into prepared mixture useful for the construction of the composite material.

Parameter	Unit	KP/PPLP	EP/PLS	EP/PPLP
SiO_2	% wt.	32.92	43.70	19.90
CaO	% wt.	49.03	12.14	35.70
MgO	% wt.	3.50	1.91	1.64
Al_2O_3	% wt.	14.13	20.45	10.56
Fe_2O_3	% wt.	0.82	5.23	2.20
MnO	% wt.	0.05	0.10	0.03
K_2O	% wt.	0.55	2.95	0.40
P_2O_5	% wt.	0.31	0.46	0.14

Table 5.
Average composition of ash [26].

The process is based on mixing wet biodegradable sludge with ash where the ash is a wasted by-product of energy generation. In general, the rule is that the ash produced from paper, paper sludge, wood chips, and wood biomass (PPLP) contains more live lime (CaO) than ash resulting from the incineration of coal, lignite, and peat (PLS). The boiler dust (KP) and the electric filter ash (EP) are suitable for the hygiene and stabilization of the BS at the CEN-EN 12832, where KP/PPLP is more effective than the EP due to a higher content of free CaO (**Table 5**).

After mixing, chemical hydrolyzation and hydration of active earth metal and metal oxides (CaO, MgO, Fe_2O_3) present in ash occurs:

$$Me\,(H_2O)_n{}^{m+} + H_2O \rightleftarrows Me\,(H_2O)^{(m-1)+}\,(OH)^{(m-1)+} + H_3O^+ \quad (1)$$

$$n\,Me^{m+} + m\,OH^- \rightarrow Me_n\,(OH)_m + \Delta H_h \quad (2)$$

This reaction releases heat, which increases mixture temperature up to 65°C and pH value above 12. Under that conditions heavy metals convert into heavy soluble metal hydroxides $Me_m(OH)_n$. However, different heavy metal hydroxides have different solubility depending on pH (see **Figure 1**), and adjustment of pH of the mixture is necessary to get chemically inert product. Due to high pH value of the mixture that can increase up to 12.4, the salts of heavy metals are converted into water-insoluble form which prevents further extraction.

Because of alkaline conditions, proteins in the presence of water undergo hydrolytic decomposition and ammonia is formed. In such conditions all pathogenic microorganisms and their spores are destroyed which ensures biological and biochemical stability of the product. A few minutes after stabilization, no *Salmonella* was detected, and the number of *Escherichia coli* was below legal limits [27].

After that the mixture passes into the phase of solidification. Because of a specific mixture of oxides and the CaO content, ash has a high pozzolanic power comparable to Portland cement. Pozzolans are a mixture of silicate and aluminum oxides which itself do not possess the cement values but, in dusty form and in the presence of water, react with CaO already at room temperature and form cement-like materials. The alkali conditions trigger a set of pozzolanic reactions similar as in crystallization of the cement.

Using X-ray diffraction spectrometry lime, portlandite, calcite, quartz, alumina, muscovite, cellulose, and other different C-H-S structures were detected (see **Figure 2**) [29]:

$$CaO + H_2O \rightarrow Ca(OH)_2 + \Delta H_h \qquad (3)$$

$$3Ca(OH)_2 + 2SiO_2 \rightarrow 3(CaO)_2(SiO)_2{\cdot}3(H_2O) + \text{C-S-H structures} \qquad (4)$$

$$3Ca(OH)_2 + Al_2O_3 + 3H_2O \rightarrow 3(CaO)Al_2O_3{\cdot}6(H_2O) \qquad (5)$$

After 28 days, ettringite structures $Ca_6Al_2(SO_4)_3(OH)_{12} \times 26H_2O$ may be developed in the presence of sulfates.

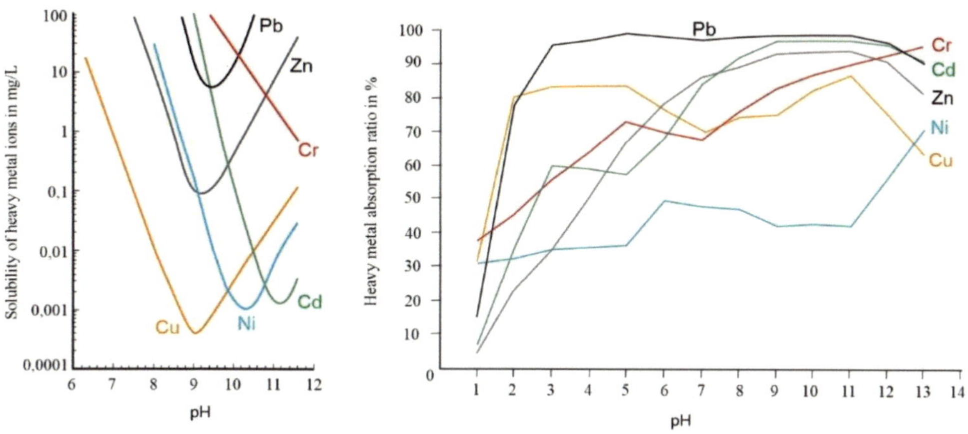

Figure 1.
Solubility of heavy metal in independence of pH (left) and adsorption of heavy metal ions in dependence of pH (right).

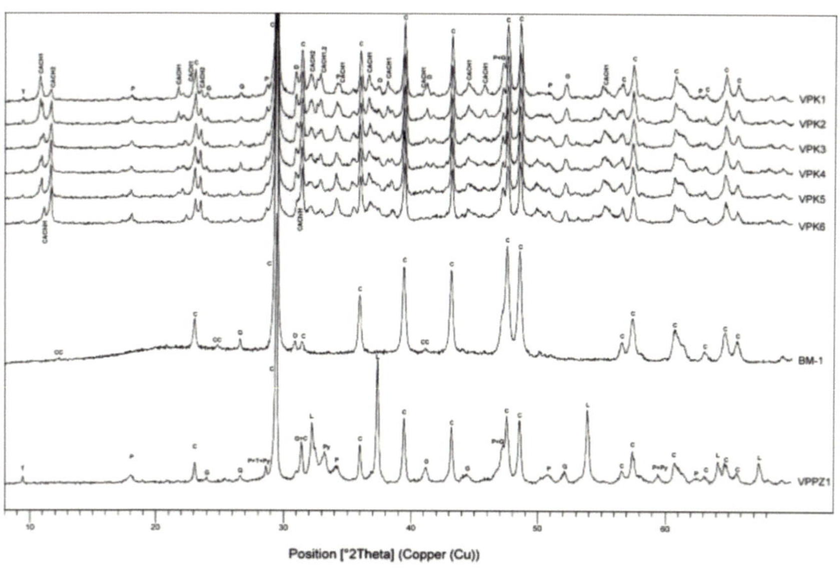

Figure 2.
*X-ray diffraction pattern of biodegradable sludge (BM), biomass ash (VPPZ1), and the composites after 3, 7, 14, 28, 56, and 90 days (VPK-1 to VPK-6, respectively). Legend: C, calcite; T, talc; P, portlandite; Q, quartz; L, lime; G, gehlenite; CC, clinochlor; D, dolomite; Py, pyrite; CACH1, $Ca_8Al_4O_{14}CO_2$*$24H_2O$; CACH2, $Ca_4Al_2O_{14}CO_9$*$11H_2O$; CaCh1H, $Ca_4Al_2O_6Cl_2$ *$10H_2O$ [28].*

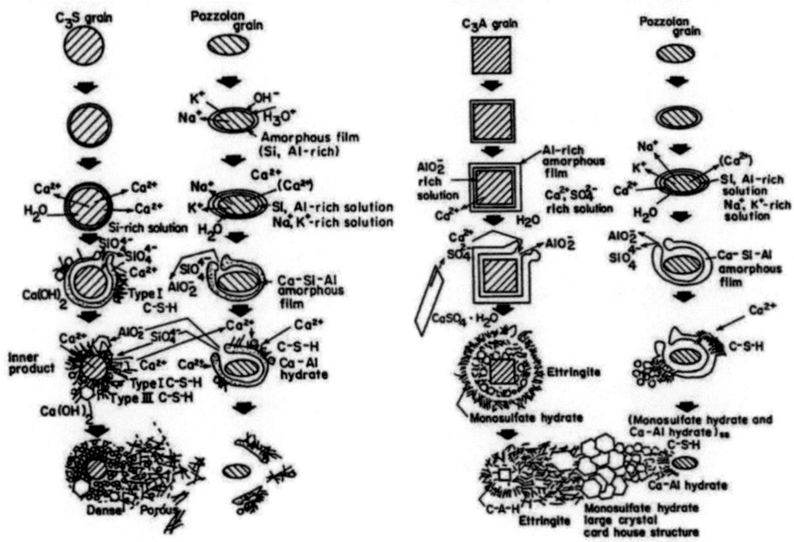

Figure 3.
Schematic explanation of the mechanism of hydration in hydration in the C3S-pozzolan system and the C3A-pozzolan system in the presence of $Ca(OH)_2$ and $CaSO_4\,2H_2O$ [30].

Component	Detection method	Legal limit in leachate in mg/kg s.s.	Result in mg/kg s.s.
As	SIST EN ISO 17294-2:2005	0.5	<0.02
Ba	SIST EN ISO 17294-2:2005	20	0.94
Cd	SIST EN ISO 17294-2:2005	0.04	<0.005
Cr (total)	SIST EN ISO 17294-2:2005	0.5	<0.01
Cu	SIST EN ISO 17294-2:2005	2	0.94
Hg	SIST EN ISO 17294-2:2005	0.01	<0.001
Mo	SIST EN ISO 17294-2:2005	0.5	<0.05
Ni	SIST EN ISO 17294-2:2005	0.4	0.18
Pb	SIST EN ISO 17294-2:2005	0.5	0.15
Sb	SIST EN ISO 17294-2:2005	0.06	0.013
Se	SIST EN ISO 17294-2:2005	0.1	<0.01
Zn	SIST EN ISO 17294-2:2005	4	0.21
Chloride	SIST EN ISO 10304-1:2009	800	57.3
Fluoride	SIST EN ISO 10304-1:2009	10	1.2
Sulfate	SIST EN ISO 10304-1:2009	6000	25.7

Table 6.
Main characteristics of leachate of final product produced by mixing BS and pozzolanic ash in ration 70/30 [29].

Crystal structures (see **Figure 3**) capture heavy metal hydroxides, other pollutants, and stabilized organics inside the net and prevent further extraction of pollutants. At the same time, crystal structures give the final material geomechanical characteristics (**Tables 6** and 7).

1.2 BACOM technology

BACOM technology is an approved technology for the processing of BS in construction composites. It is based on the idea of the alkalization of biological sludge by mixing with the ash, which expresses pozzolanic activity. The BACOM technology operates in

Property	Measuring method	Result
Humidity	ISO/TS 17892-1:2001/AC:2010	23% wt.
Max. dry density	SIST EN 13286-2:2010/AC:2013	1.2 Mg/m^3
Optimal humidity on standard Proctor test	SIST EN 13286-2:2010/AC:2013	29.1% wt.
Uniaxial compressive strength	SIST EN 13286-41:2004	96 kPa
Shear strength	SIST TS ISO/TS 17892-10:2004/AC:2010	φ = 38.47 c = 42.6 kPa
Compressibility module at load rate: • 50 kPa • 100 kPa • 200 kPa • 400 kPa • 800 kPa	SIST TS ISO/TS 17892-5:2004/AC:2010	8213 kPa 8213 kPa 10,704 kPa 15,850 kPa 19,390 kPa 23,693 kPa
Water impermeability at load rate 200 cm/s	SIST TS ISO/TS 17892-11:2004/AC:2010	2.35 × 10^{-7}

Table 7.
The main geomechanical characteristics of final product produced by mixing BS and pozzolanic ash in ration 70/30 [29].

accordance with the European CEN-EN 12832 and complies to the conditions for the processing and use of biodegradable, municipal, and similar sludge, including chemical hygienization and inertization of BS and by mixing with live lime and/or ash.

It enables to mix biodegradable sludge, which contains from 2 to 30% of dry matter with ash (or other alkaline materials) which usually contains up to 80% of earth alkali and/or metal oxides. When the content of dry matter increases up to 60%, the mixture passes to semisolid state and solidify after a short time (in average after 72 hours). The construction and mechanical characteristics can be improved by further admixing different materials such as ash, cement, lime, micro silica, porcelain, slag, foundry sand, natural and artificial fibers, and different kinds of vermiculites. With careful choice of additives, the chemical, mechanical, geotechnical, and hydromechanical characteristics of produced materials can be adjusted and improved before further application.

The technological process of BACOM includes three main process operations that are shown in **Figure 4**:

- In the first stage, raw biodegradable sludge (1) is mixed with ash (2).
- Inside mixing device (3), semisolid mixture (4) is formed.
- Thickened material (4) is additionally admixed with composite material (5) and the final product (6) is formed.

Applying the BACOM technology in the processes of disposal of the biodegradable sludge enables reduction of investment costs in the beginning due to the smaller storage capacities needed for storage biodegradable sludge and alkaline materials. Due to the fast exploitation of the biodegradable sludge, the biological decomposition of the organic components is reduced on the lowest possible level, and because of that, additional reduction of negative impacts on the environment is achieved. The technology can be built as independent facility for the processing of the biodegradable sludge or as technological part of the existing biological wastewater treatment plant.

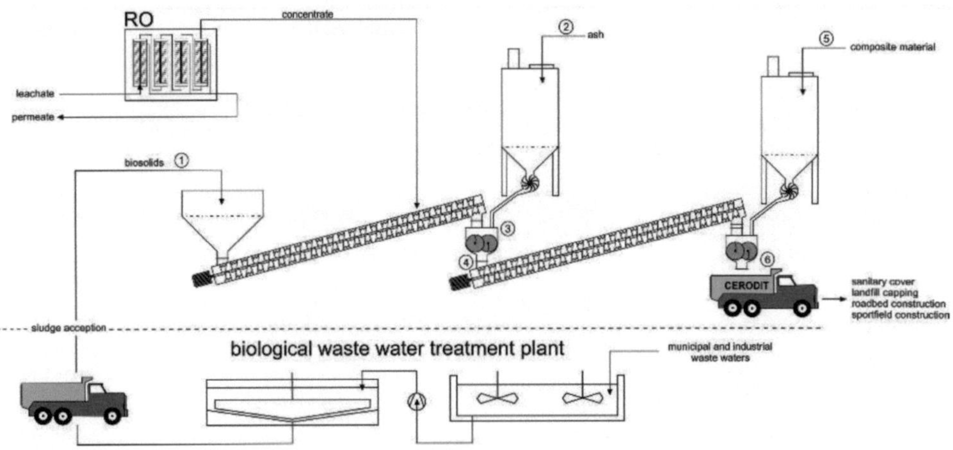

Figure 4.
Schematic explanation of BACOM process.

Overall, the implementation of BACOM technology into processes of recycling and disposal of BS can decrease investment costs for about 90% comparing to the other mentioned technologies, and overall costs of disposal calculated on dry matter can decrease by 88%. At the same time, the greenhouse effect decreases by 95% compared to incineration. The produced composite can be of further use for cheap replacement for geo-composite material or clay which additionally contributes to decreasing of the greenhouse effect due to reduction of land degradation which is a consequence of the opening of a new mining site.

Up to today the usage of alkaline waste for the stabilization of biodegradable sludge did not get application in big extent due to batch processes of stabilization which demands organization of relatively large storage capacity for biodegradable sludge and large storage capacities for final product. During the uncontrolled storage of the sludge, aerobic and anaerobic degradation processes occur inside the sludge that cause organic decay of organic part and consequently releasing of greenhouse gasses and unpleasant smell into environment. The BACOM technology solved those issues with enabling continuous stabilization of biodegradable sludge with online mixing of the alkaline materials and composite materials into the sludge, and the production of construction composite materials and long storage is not needed. The solution is based on innovative connection of two batch processes inside one efficient continued process, which at the end leads to sustainable production of replacements of geo-composites and clay.

2. Case study

The example for the determination of sustainability of BACOM technology compared to other available technologies has been placed in Slovenia, which represents a small, closed, and relatively densely populated area. Slovenia has a smaller incinerator in Celje, so the incineration of BS was not taken into account in the performance of the study, but we have therefore studied possible involvement of Eco-Bis technology. The study model is appropriate for smaller, more populated areas with less developed systems for BS management. According to the ARSO (Slovenian Environment Agency) data, in Slovenia 203,059 tons of BS or approximately 100,000 tons of dry matter was produced in 2017. About 70,000 tons of these were exported to Hungary for the preparation of artificial soil; 34,000 tons were driven to incineration; 12,000 tons in landings; 20,000 tons on composting; and the rest wastes were exported to other processing operations. The use in agriculture has been exceptionally small already from 2006. The main transport distance for the transport of BS to processing/landings was less than 250 km.

3. Methodology

Environmental profiles and comparative studies were made by the LCA methodology standardized according to ISO 14040 and 14,044, which is suitable for comparison and evaluation of different technologies within the prescribed boundaries. Using the SimaPro 7.1 software package and its database, the comparison was made by the IMPACT 2002 + method [31]. The methodology included (1) purpose and definitions, (2) inventory list, (3) determination of effects, and (4) interpretation. The specifics of the methodology, the main hypothesis, the key assumptions, and conclusions will be described through the accompanying text.

3.1 Purpose and definition

The aim of the study was to evaluate and compare the different management systems with BS in closed densely populated areas, as well as defining environmental impacts and energy balances. Although in most cases the transport of BS represented a large proportion of emissions, transport was neglected with the reason that the boundaries were determined at the entrance to the processing. The independent evaluation of individual technologies was enabled. In the analysis, due to the continuation of the calculation, the drying process of BS from 2 to 20% of the dry matter is excluded. All calculations were based on the presumption that biodegradable sludge with 20% of dry matter is being processed. The LCA also ignored the construction and dissembling of individual technologies because in this analysis, only the influence of the working activities of individual technologies is of interest.

Scenario 1 (landfilling) represents the disposal of BS on sanitary landfills which have arranged system for capturing biogas and burning of biogas on torches. According to the predictions, about 45% of the generating biogas is captured, and 55% of the generated biogas is emitted into the atmosphere through the different part of the landfill as, for example, drainage system and the boundary slopes. Although such a scenario is undesirable and banned in the EU, it should be considered due to undeveloped and inefficient management systems and because it is used in more than 50% of examples.

Scenario 2 (composting) represents composting of BS with 20% of dry substance, with a technique of aerobic digestion in open digs equipped with active aerating systems. According to the experience and literary data, it was presumed that active composting lasts 24 days and further ripening of the compost for another 60 days. The compost with approximately 40% of dry substance is landfilled on sanitary depot. Power consumption for processing 1 ton of BS in compost is 90 kWh; embedding of the compost in the body of the landfill requires additional use of 0.6 kg of diesel fuel for 1 ton of BS.

Scenario 3 (AD&L) represents anaerobic digestion of BS and use of a manufactured biogas for electricity production. The heat is not used, and the compost is embedded in the sanitary landfill. For anaerobic processing 1 ton of BS with 20% of dry substance in average 34 kWh of electric power and 20 liters of water is used. Processing of BS with 20% of dry matter in biogas and its active use ensures production of about 175 kWh surplus of electricity.

Scenario 4 (BACOM) represents the solidification/stabilization of the BS with pozzolanic ash in proportions from 30/70 to 70/30 which depends on the final use of the product. The product can be used for replacing clay or bentonite. For the processing of 1 ton of BS with 20% of dry matter, 0.3 kWh of electric power and additional 0.6 kg of diesel fuel for its embedding are used. The emissions into the environment are reduced to 1% compared to uncontrollable landfilling of BS. From 1 ton of BS with 20% dry matter, up to 1.3 tons of clay replacement can be produced.

Scenario 5 (Greenlife Eco-Bis) represents a technology for controlled pyrolyzing of BS into biochar. BS is dried from 20 to 75% dryness with the use of the heat produced with pyrolytic processes. The average use of energy for drying is 192.73 MJ/kg of BS. After that, the dry sludge is pyrolyzed into bio-coal under controlled conditions. The gain of heat using pyrolytic processes is around 243.10 MJ/kg of dry BS, taken into account that the electricity consumption for processes itself is 40 kWh. From 1 ton of BS with 20% of dry substance, the 300 kg of coal can be produced. Produced biochar can efficiently replace artificial and growing fertilizers in agriculture.

3.2 System boundaries

System boundaries are defined with the input of BS at the entrance into the processing. Due to the equalized assessment, transport and system of drying of the BS are neglected. It is necessary to note that the technology of the BACOM drying the sludge is not necessary because it also works with the sludge where the content of the dry substance is lower than 10%. Greenlife Eco-Bis technology has a built-in effective drying system with the system of vacuum filtration which is exploit surplus heat which is generated within pyrolytic processes. System boundaries are graphically displayed in **Figure 5**.

3.3 Functional unit

The functional unit is used for the definition of input or output from the system. The purpose of introducing a functional unit is equalizing evaluation of different scenarios. In our case, the functional unit used for evaluation is 1 ton of BS containing 20% of dry substance.

3.4 Inventory analysis

At this stage, the input materials, energy used, as well as emissions into the atmosphere, water, and soil were evaluated. Data about the BS which is processed by the

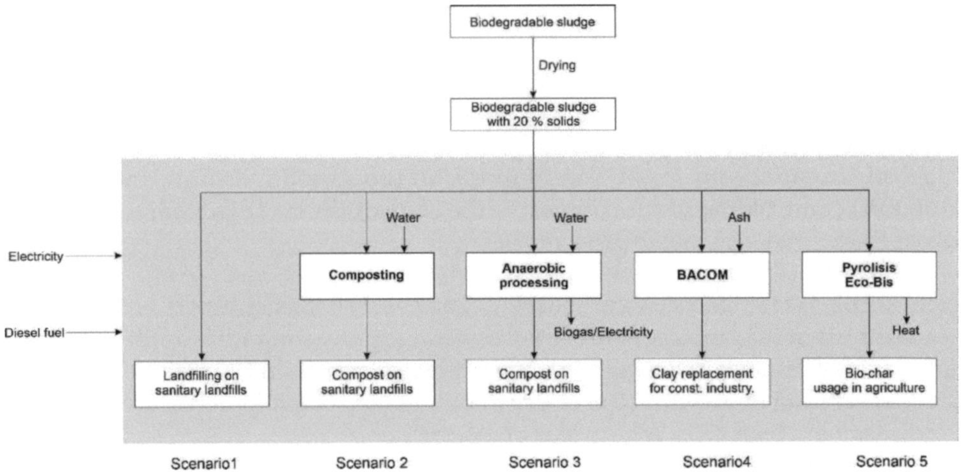

Figure 5.
System boundaries for LCA.

	Landfilling	Composting	Anaerobic	BACOM	Pyrolysis
Scenario	**1**	**2**	**3**	**4**	**5**
Emissions into air					
CO_2	14.23	61.81	61.81	0.1423	0.1020
CO	0.02	0.10	0.10	0.0020	0.0128
CH_4	15.14	0.05	—	0.1514	0.0002
VOC	0.02	0.01	0.30	0.0002	—
NO_X	0.04	—	0.30	0.0004	—
Nitrogen	4.63	—	—	0.0463	—
Oxygen	0.02	—	—	0.0002	—
H_2S	0.02	0.02	—	0.0020	—
PM_{10}	—	0.11	—	—	—
NH_3	—	0.04	—	—	—
SO_2	—	0.01	0.01	—	—
HCl	—	0.01	0.01	—	—
Emissions into water					
Chlorides	0.028	0.028	0.028	0.00028	0.1020
KPK_5	0.011	0.011	0.011	0.00011	0.0128
BPK_5	0.056	0.056	0.056	0.00056	0.0002
Nitrogen	0.003	0.003	0.003	0.00003	0.0062
Suspended part.	0.002	0.002	0.002	0.00002	0.0146
NH_4OH	0.001	0.001	0.001	0.00001	0.0005
P (total)	0.001	0.001	0.001	0.00001	0.0000

Table 8.
Partial list of emissions into the atmosphere and water for particular scenarios in kg per ton BS with 20% dry solids.

company CeROD d.o.o. and CeROP (for year 2013 and 2015) were obtained from the BS analysis; data on emissions of gasses in the process of degradation of BS were measured and equalized with data obtained from the peer reviewed literature [14, 15, 32] and are shown in **Table 8**. Data about electricity production for Slovenia were obtained from the database BUWAL 250 and data about clay from the ETH-ESU database.

4. Discussion

Potential effects of five different scenarios (technologies) are listed in **Tables 9** and **10**.

4.1 Impacts on human health

From **Table 10** and **Figure 6**, it is evident that the total impact on human health in the case of landfilling of BS, which is equipped with the system for the active capture and incineration of biogas, is by 16.55 units lesser than the impact on human health caused by composting. Such conclusions come from the fact that the effects in the case of landfilling are limited to fenced and guarded spaces and that there are no emissions of dust, as well as from the fact that the disposal of BS does not require additional

Subgroup	Unit	Landfill	Compost	AD&L	BACOM	Eco-Bis
Emissions of carcinogenic substances	DALY	3.03E-05	0.000613	0.000231	−5.08814	1.75173
Emissions of non-carcinogenic substances	DALY	0.024072	0.032287	0.032183	−4.2439	−11.8864
Emissions of substances harmful for respiratory system	DALY	5.046047	21.85825	21.85621	−291.066	−162.069
Ionization	DALY	1.66E-07	1.66E-07	0	−0.38479	−0.00135
Emission of substances which destroying ozone layer	DALY	4.73E-07	8.39E-07	1.45E-07	−0.35543	−0.0001
Emission of organic pollutants harmful for respiratory path	DALY	0.291465	0.018953	0.018951	−0.72645	0.007838
Aquatic toxicity	PDF*m2*l	2.68E-05	0.000227	9.92E-05	−0.92158	−1.75714
Soil toxicity	PDF*m2*l	2.1E-05	0.002301	0.002271	−21.251	−0.24776
Soil acidification	PDF*m2*l	0.166662	0.454253	0.454224	−3.56515	−3.98712
Land use	PDF*m2*l	5.86E-07	5.86E-07	0.00	−5.03381	−0.33184
Acidification of water environ.		—	—	—	—	—
Water eutrophic.		—	—	—	—	—
Climate change	kg CO2(eq)	121.444	0.357449	0.35498	−176.323	−203.961
Non-renewable energy sources	MJ prim.	0.000434	0.005132	0.001863	−162.77	−212.651
Extraction of minerals	MJ prim.	1.52E-08	1.52E-08	0.00	−1.3E-05	−0.0015

Table 9.
List of effects on subgroups IMPACT 2002+.

Effect on:	Unit	Landfill	Compost	AD&L	BACOM	Eco-Bis
Human health	DALY	5.361614	21.91011	21.90758	−301.864	−172.197
Ecosystem	PDF*m^{2}*l	0.166711	0.456783	0.456594	−30.7715	−6.32386
Climate changes	kg $CO_{2(eq)}$	121.444	0.357449	0.35498	−176.323	−203.961
Natural resources	MJ prim.	0.000434	0.005132	0.001863	−162.77	−212.652

Table 10.
List of effects on groups IMPACT 2002+.

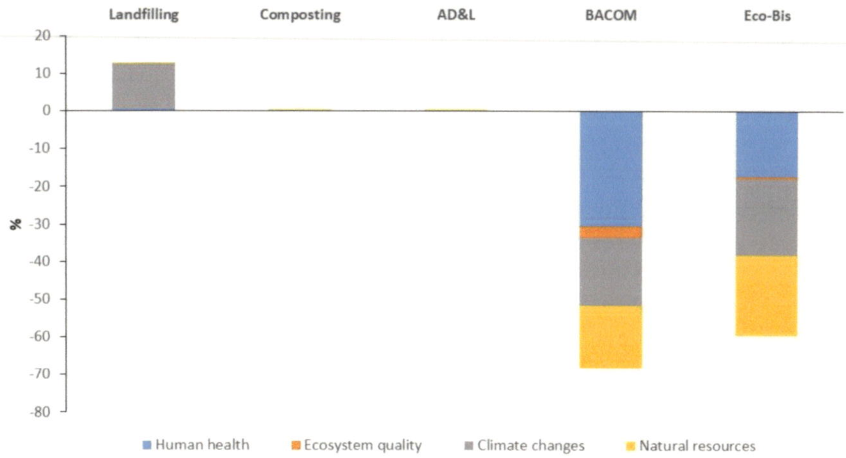

Figure 6.
Overall LCA result.

energy consumption for processing. The reduction of the impact on human health by additional 166.84 units is reflected in the use of Eco-Bis technology instead of landfilling of BS primarily due to the energy utilization and consequentially direct emission reductions as well as additional emissions reductions because of the replacement of artificial fertilizers with environmental friendly bio-coal. The additional reduction in impacts by the 135 units was calculated in case of using the technology of converting BS into replacements for clay and bentonite geo-composites (BACOM). Opening of new clay mines, where energy demanding processes are applied for clay production are not needed anymore in the case of implementation of BACOM technology.

4.2 Impacts on ecosystems

From **Table 10** and **Figure 6**, it is evident that direct landfilling of BS on the waste disposal site has approximately 2.5 times higher negative impact on the ecosystem as a controlled aerobic or anaerobic digestion of BS which also includes disposal of the produced compost on sanitary landfill. This is the result of the reduction of uncontrolled emission of biogas and toxic substances into the environment. A decrease of additional 6.16 units was calculated in the case of the use of Eco-Bis instead of landfilling of BS due to the prevention of its biological decomposing where decreasing impacts on the environment are based primarily on the replacement of the artificial fertilizer with biochar. Further reduction of negative impacts on the ecosystem for 24.45 units can be achieved by using BACOM technology and is the result of prevention of biologic breakdown of BS and of the fact that the manufactured material can be used as a substitute for clay, bentonite, and plastic materials.

4.3 Impacts on climate change

From **Table 10** and **Figure 6**, it is evident that landfilling of BS on the waste disposal site has 344 times more negative impact on climate change than composting and anaerobic processing and the subsequent disposal of compost to the waste disposal site. This is the result of the reduction of uncontrolled emission of greenhouse gasses into the environment. However, additional reduction of 297.76 units was noted when using BACOM technology which is the result of prevention of biologic breakdown of BS and the fact that the manufactured material can be used as a substitute for clay, bentonite, and plastic materials. An additional reduction of 27.6 units was calculated in the use of Eco-Bis technology, and it is the result of a completely prevented biological degradation of BS and of the replacement of the artificial fertilizers with biochar.

4.4 Impacts on the use of natural resources

From **Table 10** and **Figure 6**, it is evident that a landfilling of BS on sanitary landfill, composting, and anaerobic digestion and the subsequent disposal of the compost onto the landfill have no impact on the use of natural resources. Additional reduction of 162 units is enabled with the implementation of the BACOM technology because the manufactured material can be used as a substitute for clay, bentonite, and in some cases as plastic. The additional reduction of 66.9 units is enabled by the usage of Eco-Bis charcoal as the replacement of artificial fertilizers (**Table 9** and **Figure 6**).

The overall sustainability of using different technologies from the highest to the least efficient is as follows: BACOM > Eco-Bis > AD&L > composting > disposal.

5. Conclusions

According to expectations, landfilling of BS in sanitary landfills is the least acceptable option, even though modern landfills are equipped with modern biogas capture systems. Composting is an acceptable and widely accepted option because it is cheap and has rather neutral effect on the sustainability of the management of BS. However, considering the investment and operational costs which have no economic effect, it is less desirable option than anaerobic digesting, where the produced biogas can be exploited for energy production.

All abovementioned solutions need centralized organization. Successful operation requires large surfaces, and due to that, the local community must agree with the implementation of such a plant inside local areas. In addition, risk management in centrally organized technologies requires extensive and complex logistic operation and relatively large operating costs to achieve a small economic gain.

Material processing and material use of BS seems to be a much more acceptable and sustainable option than landfilling or incineration, because cheap replacements for materials produced from nonrenewable sources (e.g., artificial fertilizers, clays, bentonite, or even plastics in some cases) can be produced. In many cases these processes reduce negative pressures on the environment and improve life in the local communities.

These installations are usually small and mobile and can be placed directly in the vicinity of the WWTPs. The problems with burdening of the environment and rising costs because of extended logistics are solved or at least minimized in such a case. In addition, material processing and use of manufactured material mean the production of products with an added value, which can be used as raw materials or semifinished products in other industrial sectors, what is in accordance with the principles of industrial eco symbiosis and circular economy. The aforementioned technologies enable the creation of new jobs and the reduction of wastewater treatment costs.

The involvement of material processing technologies is in a consensus with the European Directive on waste (2008/98/EC) as shown in **Figure 2**.

Each technology has its own benefits or deficiencies, but in general the central organizing technologies as, for example, incineration, anaerobic digestion, composting, and landfilling, are more appropriate for processing of BS in a bigger scale (e.g., quantities above 30.000 tons of biosolids per year); meanwhile, smaller and more flexible technologies for material processing are more suitable for processing of BS in quantities below 30,000 tons per year.

Author details

Marko Likon[1*] and Marjan Zemljič[2]

1 Škrlj d.o.o., Slovenia

2 Ekorec d.o.o., Slovenia

*Address all correspondence to: marko.likon@telemach.net

References

[1] Fytili D, Zabaniotou A. Utilization of sewage sludge in EU application of old and new methods—A review. Renewable Sustainable Energy Rewiews. 2006;**12**(1):116-140

[2] Kroiss H. What is the potential for utilizing the resources in sludge. Water Science and Technology. 2004;**49**(10):1-10

[3] Aubain P, Gazzo A, Le Moux J, Mugnier E, Brunet H, Landrea B. Disposal and Recycling Routes for Sewage Sludge. Luxembourg: EC; 2001. p. 38. ISBN: 92-894-1798-6

[4] LeBlanc RJ, Matthews P, Richard RP. Global Atlas of Excreta, Wastewater Sludge, and Biosolids Management: Moving Forward the Sustainable and Welcome Uses of a Global Resource. Nairobi, Kenya: United Nations Human Settlements Programme (UN-HABITAT); 2008. p. 632

[5] EPA. Biosolids Generation, Use, and Disposal in the United States. Washington, DC: EPA; 1999. p. 81

[6] ADB. Promoting Beneficial Sewage Sludge Utilization in the People's Republic of China. Mandaluyong City, Metro Manila, Philippines: DBA; 2012. p. 33

[7] Manfredi S, Pant R. Improving the environmental performance of bio-waste management with life cycle thinking (LCT) and life cycle assessment (LCA). The International Journal of Life Cycle Assessment. 2013;**18**:285-291

[8] Milieu Ltd, WRc, RPA. Environmental, Economic and Social Impacts of the Use of Sewage Sludge on Land. Final Report. Part III. Projects Interim Reports. Belgium: EU; 2010. p. 33

[9] Eurostat. Environmental Statistics and Accounts in Europe. Luxembourg: EC; 2010. p. 346

[10] EC. Communication from the Commission to the Council and the European Parliament on Future Steps in Bio-Waste Management in the European Union. Luxembourg: EC; 2010. p. 235

[11] Arthurson V. Proper sanitization of sewage sludge: A critical issue for a sustainable society. Applied and Environmental Microbiology. 2008;**74**(17):5267-5275

[12] Sonesson U, Björklund A, Carlsson M, Dalemo M. Environmental and economic analysis of management systems for biodegradable waste. Resources, Conservation and Recycling. 2000;**28**:29-53

[13] Peters GM, Rowley HV. Environment comparison of biosolids management systems using life cycle assessment. Environmental Science & Technology. 2009;**43**:2674-2679

[14] Righi S, Olivieto L, Pedrini M, Buscaroli A, Casa CD. Life cycle assessment of management for sewage sludge and food waste: Centralized and decentralized approaches. Journal of Cleaner Production. 2013;**44**:8-17

[15] Liu Q, Jiang P, Zhao J, Zhang B, Bian H, Qian G. Life cycle assessment of an industrial symbiosis based on energy recovery from dried sludge and used oil. Journal of Cleaner Production. 2011;**19**:1700-1708

[16] Nadziakiewicz J, Koziol M. Co-combustion of sludge with coal. Applied Energy. 2003;**75**(3–4):239-248

[17] Wang N, Shih CH, Chiueh PT, Huang YF. Environmental effects of sewage sludge carbonization and other treatment alternatives. Energies. 2013;**1**:871-883

[18] Hospido A, Moriera MT, Martin M, Rigola M, Feijoo G. Environmental

evaluation of different treatment processes for sludge from urban wastewater treatments: Anaerobic digestion versus thermal processes. International Journal of Life Cycle Assessment. 2005;**10**:336-345

[19] Valderrama C, Granados R, Cortina JL, Gasol CM, Guillem M, Josa A. Comparative LCA of sewage sludge valorisation as both fuel and raw material substitute in clinker production. Journal of Cleaner Production. 2013;**51**:205-213

[20] Onaka T. Sewage can make Portland cement: A new technology for ultimate reuse of sewage sludge. Water Science and Technology. 2000;**41**(8):93-98

[21] European Commission. Green Paper on the Management of Bio-Waste in the European Union. Luxembourg: EC; 2008. p. 811

[22] ORBIT/ECN. Compost production and use in the EU, Final report of ORBIT e.V./European Compost Network ECN to European Commission. Joint Research Centre. 2008. p. 182

[23] California Integrated Waste Management Board. Current Anaerobic Digestion Technologies Used for Treatment of Municipal Organic Solid Waste [Internet]. 2019. Available from: http://www.calrecycle.ca.gov/publications/

[24] Greenlife GmbH. Pyrolisis - Recycling Sewage Sludge to Fertilizer [Internet]. 2019. Available from: www.greenlife.co.at

[25] Likon M, Černec F, Salobir M, Podobnik E. Procedure and installation for continuous stabilization and conversion of biodegradable sludge into construction composite materials. SI-Patent No. 22951. 2010. p. 10

[26] Riehl A, Elsass F, Duplay J, Huber F, Trautmann M. Changes in soil properties in a fluvisol (calcaric) amended with coal fly ash. Geoderma. 2010;**155**:67-74

[27] Retelj M. Test Report, num. P749/15-420-1. National Laboratory of Health, Food, Environment and Food, Center for Microbiological Analytics of Food, Water and Other Samples from Environment. 2015. p. 4

[28] Pavšič P, Oštir D, Mladenovič A, Kramar S, Dolenec M, Bukovec P. Sewage-sludge stabilization with biomass ash. Materials Technology. 2013;**47**:349-352

[29] Mladenovič A, Legat A. Test Report, num. V 0884/13-250-1. Slovenian National Building and Civil Engineering Institute. 2013. p. 5

[30] Ramezanianpour AA. Natural Pozzolans. Cement Replacement Materials: Properties, Durability, Sustainability. Berlin: Springer; 2014:1-46

[31] Jolliet O, Margni M, Charles R, Humbert S, Payet J, Rebitzer G, et al. A New Life Cycle impact assessment methodology. International Journal of Life Cycle Assessment. 2003;**8**(6):324-330

[32] Murray A, Horvath A, Nelson KL. Hybrid life-cycle environmental and cost inventory of sewage sludge treatment and end-use scenarios: A case study from China. Environmental Science & Technology. 2008;**19**:3163-3169

Chapter 2

Pre-treatment Technologies to Enhance Anaerobic Digestion

Sridhar Pilli, Ashutosh Kumar Pandey, Ankur Katiyar, Kritika Pandey and Rajeshwar Dayal Tyagi

Abstract

Sustainable energy production is the major priority in the world due to global warming, climate change, and fossil fuels depletion. Anaerobic digestion (AD) of sludge is the sustainable process producing the energy and minimizing the fossil fuel usage. However, conventional AD of sludge is not sustainable since it takes longer time for digestion which increases the energy input and greenhouse emissions. Therefore, pretreatment technologies have emerged to enhance methane production and thus the energy output from the AD process. In this chapter, pre-treatment technologies adopted mainly physical, chemical, thermal, and other advanced processes to enhance methane production in the last decade are elaborated. In addition, energy balance of the process and the feasibility of the pre-treatment technologies and their current status are discussed.

Keywords: waste activated sludge, pre-treatment, anaerobic digestion, methane, sustainability

1. Introduction

The production of inevitable waste activated sludge (WAS) as a by-product during the biological wastewater treatment demands for sustainable treatment options that assist in the proper utilization of sludge before its disposal. The composition of sludge mainly includes microbial cells and organic components such as proteins, carbohydrates, and lipids. Understanding the properties of the sludge may help in processing it to produce beneficial products or as a feedstock for bioenergy generation.

Anaerobic digestion (AD) for sludge stabilization is operated in about 38% of the total treatment plants, whereas only 6% of plants employ aerobic digestion and composting [1]. AD of sludge produces biogas, which mainly contains methane and carbon dioxide. AD of sludge is a sustainable process as it recovers energy from the biogas and replaces fossil fuel usage and minimizes GHG emissions. However, the efficacy of these processes is limited by the presence of complex structural components, extracellular polymeric substances, and rate-limiting cell lysis in WAS [2]. Moreover, the effects of hydrolytic enzymes are reduced in WAS as their penetration inside the bacterial cell is hindered by the cell walls, making the degradation of intracellular organic compounds tedious. Thus, it increases the digestion time and energy required for digestion processes. Hence, to overcome these drawbacks, pre-treatment technologies are adopted to break the cells and to liberate the cell constituents.

Various physical, chemical, and biological pre-treatment methods have been reported in the literature which is used individually or in combination for pre-treatment of WAS. These include treatment by hydrolysis, ultrasound, enzymatic lysis, acidification, alkaline hydrolysis, alkaline-thermal, and thermal-H_2O_2, microwave alkaline, and others [3]. The use of appropriate pre-treatment strategy can enhance the degradation and disintegration of both extracellular and intracellular substances reducing the retention time needed by biological digestion processes [4].

This chapter presents literature about the different methods of pre-treatment that have been used for enhancing the AD. Moreover, the factors affecting their operational efficiency have also been discussed. Furthermore, a brief account of the large-scale feasibility and economic aspects of the overall pre-treatment processes is discussed.

2. Pretreatment technologies

The pre-treatment technologies enable the cells constituents easily available for the microorganisms to produce the biogas. Various pre-treatment technologies (mainly mechanical, chemical, biological, and physio-chemical) and their effect on enhancing the AD and methane production presented in the literature during last decade are discussed here.

2.1 Mechanical pre-treatment

2.1.1 The process involved and mode of action

Mechanical pre-treatment disintegrates and/or grinds solid particles of the substrates, thus releasing cell compounds and increasing the specific surface area. The increased surface area provides better contact among substrate and anaerobic bacteria, which enhances the AD process [5]. A larger particle radius exhibits lower chemical oxygen demand (COD) degradation and a lower methane production rate [6]. Likewise, the particle size is inversely proportional to the maximum substrate utilization rate of the anaerobic microbes [7]. Therefore, mechanical pre-treatments such as sonication, liquid shear, collision, a high-pressure homogenizer and liquefaction are conducted to reduce the substrate particle size. During sonication the electrical energy from the source is converted to mechanical vibration which then converts to cavitation. The shear forces exerted as a result of cavitation cause WAS floc dispersion and further cell disintegration releasing organic macromolecules that are further degraded into short-chain compounds [8]. The irradiation intensity, time, and temperature-induced, as a result, can impart a cumulative effect enhancing the sludge degradation [9]. The main effect of ultrasonic pre-treatment is particle size reduction at low frequency (20–40 kHz) sound waves [10]. High-frequency sound waves also cause the formation of radicals such as HO^-, H^+, which results in oxidation of solid substances [11]. Besides ultrasonic irradiation, cavitation can also be produced by venturi meter tubes under controlled conditions of liquid flow [12].

High-pressure homogenization (HPH) pre-treatment involves the use of 30 and 150 MPa pressure for 3–30 min to pressurize the heterogeneous sludge components. The homogenization occurs due to shear, that is, when the pressurized sludge is released to impact on a ring [13]. The formed cavitation induces internal energy, which disrupts the cell membranes [14]. Both electroporation and liquefaction pre-treatments cause cellular structure damage, thus the effect on the AD process is similar to maceration [15]. Barjebruch and Kopplow treated surplus sludge with

HPH at 600 bars and showed that the filaments were completely disintegrated [14]. Increased 25% volatile solids reduction was observed in AD for HPH pretreated sewage sludge [16]. This improvement was induced by increased soluble protein, lipid, and carbohydrate concentration.

The advantages of mechanical pre-treatment include no odor generation, an easy implementation; better dewaterability of the final anaerobic residue, and moderate energy consumption. Disadvantages include no significant effect on pathogen removal and the possibility of equipment clogging or scaling [17].

2.1.2 Nanobubbles

Nanobubbles are spherical liquid structures containing gas which are stable and efficient when possessing typical overall diameters in the nanometer range (less than 10^3 nm). The presence of negative charge on nanobubbles is observed when present in pure water over a wide pH range. Nanobubbles stabilized their structures because of the same charge repulsion that occurred between adjacent nanobubbles [18]. However, some reports suggest hydrophobic attraction between negatively charged surfaces of nanobubbles and these contradictory reports could be attributed to the differences in nanobubble generation techniques, surface tension, or varying molecular arrangement at the gas-liquid interface [19]. Nanobubbles with diameters of approximately 13 nm have been well-engineered as spherical water packages with gas for food safety applications whose efficiency is well established based on bubble surface stability and the electrostatic charges present on the bubble surface [20]. Besides possessing high stability, nanobubbles in liquid systems also show a high mass transfer rate and enhanced solubility in gas [21]. Nanobubbles with a varying range of diameters have been engineered by different methods such as constant purging of octafluoropropane gas into an ultrasonicated solution of mixed surfactant which creates bubbles ranging in 400–700 nm mean diameter [22]. Palladium electrode with ultrasonication has been used to form nanobubbles of 300–500 nm diameter [23]. Nanobubbles form reactive free radicals as they collapse due to the presence of ions in groups at the gas-liquid interface [24]. The ability of nanobubbles to form reactive free radicals makes them potent applicants in the field of pre-treatment of wastewater components. In submerged systems, nanobubbles formed by the use of air or nitrogen are known to enhance the activity of aerobic and anaerobic microorganisms that improve the waste degradation efficiency and overall water quality [25]. According to the studies the higher negative charges were observed on sludge components on the addition of nitrogen gas nanobubble water the degradation of carbohydrates and proteins get increased along with methane production, that is, 29% more than that of control [25].

2.1.3 Hydrodynamic cavitation

Cavitation is a process of cavity bubble formation which burst within the liquid to create intense pressure spots and shock waves. These factors create localized energy and turbulence which causes an impact on adjacent particles and also mixing of insoluble substances like oil and water to form emulsion [26]. This mechanism is favorable in cases where AD of sludge is hindered due to the presence of lipid-containing substances which are insoluble in water. Their insolubility causes adversity in their interaction with hydrolytic bacteria which decreases the efficiency of the overall hydrolysis process. Applying localized energy supplies insignificant amounts to small elements of the liquid volume resulting in an increase of internal energy of the liquid elements to that point which causes phase change from liquid to gas and the formation of bubbles filled with vapor and gases. Following, when the bubbles leave the high energy zones, they violently implode and disappear. The localized energy could be provided by a laser beam or a stream of heavy elementary particles such as protons by molecular or optical cavitation process based on the source of applied energy [27]. Hydrodynamic cavitation was frequently been proved as a more energy-efficient method compared to other cavitation techniques [28]. Hydrodynamic cavitation for pre-treatment of sludge where cavitation was generated by using a venturi cavitation system in which bubbles are created in venturi throat (constriction) has been used. The system achieved better energy efficiency than high-speed homogenizer in terms of soluble COD/kJ WAS and also the authors observed linear relationship between total solid concentration and the increased insoluble COD for WAS indicating towards better cavitation formation at high concentration of total solids [29]. In another study, the degradation of WAS was analyzed using a novel rotation generator of hydrodynamic cavitation at pilot scale [30]. Cavitation (as a pre-treatment) of WAS resulting in an increment in soluble COD from 45 to 602 mg/L along with a 12.7% increase in biogas production due to improved AD of the pretreated WAS [30].

2.2 Thermal pre-treatment

Thermal pre-treatment of WAS has been classified as low-temperature pre-treatment (<100°C) and high-temperature thermal pre-treatment at 100–210°C [31]. High-temperature pre-treatment cause disintegration of solids in sludge, removal of pathogens at low sludge retention time and leads to biogas production but it also has been reported that exposure to high temperature causes the formation of new chemical bonds which results in agglomeration of substances present in the sludge [32]. Both the thermal pre-treatment approaches have been reported to degrade volatile solids and produce biogas, however, the efficiency of high-temperature thermal pre-treatment for solids reduction and biogas formation has been known to be comparatively higher [11]. In thermal pre-treatment strategies, temperature and time of application are the main operational parameters which decide the success of the treatment process. When effectively applied, this pre-treatment can cause the disintegration of cell membranes accompanied by the solubilization of organic compounds [33]. During WAS pre-treatment, cell wall disruption and hydrolysis due to temperature generally occur when the temperature is in the range of 160–180°C at a pressure ranging from 600 to 2500 kPa for about an hour [32].

Microwaves generate heat by causing the movement of dipoles in polar molecules, realigning them, and producing thermal effects [34]. They cause both thermal and non-thermal effects (degradation of polymeric structure) on sludge, improving biogas production and reducing volatile solids. However, microwave generation requires higher energy consumption when compared to conventional thermal pre-treatment. The increase in temperature is associated with an increase in biodegradability while a higher concentration of solids present in WAS inversely affects the degree of penetration of microwaves to the sample [35].

2.2.1 High-temperature treatment

High-temperature thermal pre-treatment is performed at temperature >100°C. The heat exchangers or direct steam injection are used to supply the steam at the desired temperature [36]. The pressure is developed as a result of steam and high temperature which is abruptly released causing a sudden pressure drop. This sudden drop in pressure along with application time and temperature comprises the major parameters necessary for efficient solubilization and subsequent methane production in AD [37]. In many cases, it has been observed that when the temperature is raised above 190°C, recalcitrant and inhibitory compounds, that is, ammonia is released which adversely affect the process [38]. Moreover, at a temperature above 180°C carbohydrates can react with protein amino terminals resulting in pyrolysis of sludge organic matter and formation non-biodegradable compounds [39]. In the case of sludge pre-treatment by high temperature, a range of 150–180°C at 600–2500 kPa @ 30 min to an hour is optimal because when the temperature is further increased, methane production is reduced due to formation of inhibitory products due to Maillard reaction. [38]. If the temperature is not in the required range, certain biomolecules are partially or incompletely degraded, e.g. proteins solubilize during sludge pre-treatment at 175°C but are not completely degraded to ammonia [40]. Some of the advantages of high-temperature thermal pre-treatment include the reduction in viscosity of sludge which in turn eases handling and transport costs [41]. Besides the reduction in viscosity, high-temperature pre-treatment at 134°C causes an increase in specific charge on sludge components as a result of colloids and extracellular polymer substance (EPS) release [42].

2.2.2 Low-temperature treatment

Low-temperature thermal pre-treatment of WAS deals with the application of <100°C temperature for a few minutes to several hours [31]. At temperatures ranging from 60 to 70°C, particle size reduction and solubilization of organic components occur [43]. Low-temperature thermal hydrolysis of sludge causes solubilization of organic matter and increase in activity of thermophilic bacteria activating the release of hydrolytic enzymes in sludge [44]. Also, rheological properties of sludge and concentration of methane in biogas during AD are positively influenced by thermal pre-treatment of sludge at low temperature [45]. The relation between pre-treatment temperature and the time of application is a very crucial factor that affects the WAS biodegradation rate [46]. It has been reported that deflocculating or reduction in the size of particles is observed when the temperature is applied in the range of 50–95°C resulting in an increased surface area which in turn increases the rate of hydrolysis in WAS [47]. The type of sludge being pretreated by thermal exposure also affects the efficiency of temperature treatment. At 70°C, the total percentage of volatile suspended solids removed from WAS was reported to be 17% which for primary non-stabilized raw sludge was only 28% indicating towards its low biodegradability [48].

2.3 Chemical pre-treatment

2.3.1 Procedure and mode of action

In chemical pre-treatment methods alkali, acid or advanced oxidation methods are used to disintegrate the organic sludge components and disrupt microbial cells (**Table 1**). AD generally requires an adjustment of the pH by increasing alkalinity, thus alkali pre-treatment is the preferred chemical method [87]. The increase in pH of WAS due to alkali pre-treatment causes many effects on sludge components which include saponification of lipid bilayer and protein denaturation in the cell membrane, solubilizing EPS by ionization of its carboxyl and amino groups and hydrolysis of sludge organic substances [99]. In the literature it was stated that excessive reagent doses can inhibit the anaerobic microbes and AD, which makes it important to control the amount and type of reagent used along with the pH desired [100]. Besides treatment with alkali and acids, oxidation processes like ozonation are also employed to increase sludge hydrolysis and biogas production rate. An advanced oxidation process like ozonation depends upon the oxidation

Type of sludge	Pre-treatment method			Biological coupling		Ref.
	Method	Condition	Outcome	Method	Outcome	
Physical approach						
Activated sludge (40.8 g TS/kg)	MW	Power = 800 W Duration = 3.5 min Energy = 336 kJ/kg TS	Increase of SCOD: 214%	AD, semi-continuous, 37°C, SRT 20 d, 42 d	+50% biogas production, +66.6% DS removal	[49]
Activated sludge (23 g TS/L)	Ultrasonication	Frequency = 24 kHz Power = 300 W Energy = ~5000 kJ/kg TS	DDcod: 9%	AD, semi-continuous, 37°C, HRT 20 d, 80 d	+35% methane yield, 0.86 energy ratio	[50]
WAS	High-pressure thermal hydrolysis	Temp = 150°C Pressure = 3 bars Duration = 30 min	36% of active, Heterotrophs converted to readily biodegradable COD and 64%	Aerobic digestion	Increase in total mass = 21%	[51]
WAS	MW	Power = 600 W Temp = 85°C Duration = 2 min	COD solubilization up to 8.5%	NR*	NR	[52]
WAS	MW-Alkali	Power = 600W Duration = 2 min	COD solubilization increased up to 46%	Aerobic digestion	Soluble COD reduction = 93% VSS reduction = 63%	[52]
Mix primary and secondary sludge	Ultrasound	Power = 480 W Duration = 30 min	Dissolution of chemical oxygen demand up to 44.4%	Thermophilic aerobic digestion	VSS removal efficiency = 55% RT = 3 d	[53]
Activated sludge	Electro kinetic disintegration	Power = 19 kV Frequency = 110 Hz Duration = 1.5 s	Increase of SCOD/TCOD: 4.5 times, increase of exocellular polymers: 6.5 times	AD, Batch, 35°C, 20–30 d	+2.5 times higher biogas production	[54]
Thickened sludge	MW	Power = 1250 W Frequency = 2450 MHz Intensity = 100%	Increase of SCOD/TCOD from 0.06 to 0.2	AD, Semi-continuous TPAD	+106% biogas production, the maximum VS removal: 53.1%	[55]

Type of sludge	Pre-treatment method			Biological coupling		Ref.
	Method	Condition	Outcome	Method	Outcome	
WAS (14.2±0.7 g TS/kg)	MW	Energy = 14,000 kJ/kg TS	Increase of SCOD/TCOD from 2 to 21%	Aerobic digestion, Batch, 35°C, 35 d	+570.7% biogas production	[56]
Meat processing Wastewater sludge	Alkaline-MW	Temp = 140°C Duration = 30min, pH = 13	Sludge disintegrate on the degree increased up to 54.9VS solubilization increased up to 42.5%	Anaerobic digestion	Increase in biogas production = 44.5%	[57]
Excess sewage sludge	Ultrasonic-Fenton	Ultrasonic density = 720 W/L Duration = 20 min Fe2 + dosage = 0.4 g/L H_2O_2 dosage = 0.50 g/L Duration = 20 min	Soluble COD increased up to 2.1 fold	NR	NR	[58]
Thickened sludge (43.6 g TS/kg)	Ultrasonication	Power = 100 W Duration = 8 min Energy = 96 kJ/kg TS	Increase of SCOD: 1741%	AD, semi-continuous, 37°C, HRT 20 d, 67 d	+27% biogas production	[59]
Thickened sludge (43.6 g TS/kg)	MW	Frequency = 2.45 GHz Power = 800 W Duration = 1 min Energy=96kJ/kg TS	Increase of SCOD: 117%	AD, Semi-continuous, 37°C, HRT 20 d, 67 d	+ 20%biogas production	[59]
Primary sludge	Electro kinetic disintegration	Energy = 33 kWh/m^3	Accumulation of acetate increased by 2.6-fold	AD, MEC, anode potential: -0.3 V vs Ag/AgCl	+2.4-fold current density (~3.1 A/m^2)	[60]
Activated sludge	Electro kinetic disintegration	Energy = ~34kWh/m^3	Increase of SCOD: 220%	AD, CSTRs, 37 ± 1°C, SRT 20 d	+33% methane production, +18% TCOD removal, –40% digester size	[61]
WAS	Visible-photocatalysis	NR	Increase COD degradation up to 61.1%	AD	Up to 7866.7 mmol H_2/L-sludge of hydrogen production was achieved	[62]

Type of sludge	Pre-treatment method			Biological coupling		Ref.
	Method	Condition	Outcome	Method	Outcome	
Mix waste activated and digested sludge	UV Photocatalysis	Catalyst = TiO_2 Duration = 4 h Temperature = 35°C UV intensity = 0.7 mW/cm^2	Soluble COD concentration increased from 1087.2 to 1451.6 mg/L for 8 h pre-treatment	AD	Methane production=1266.7 mL/L sludge, VS reduction = 67.4%, total COD reduction= 60.5%	[63]
Mixed sludge (132 ± 1 g TS/kg)	Ultrasonication	Power = 150 W Duration = 45 min	Increase of TOC: 81.5%, increase of TN: 50.0%	AD, Batch, 35°C, OLR 0.9 ± 0.1 kg VS/m^3 d	+95% methane yield	[64]
Dewatered sludge (15–20% TS)	Thermal hydrolysis	Full-scale CAMBI™ Temperature = 160°C Pressure = 6 bar	SS removal: 20–30%, increase of SCOD/TCOD from 0.04 to 0.4	AD, semi-continuous, 42 and 55°C, HRT 1–6 d, 142 d	+2–5 times in VFAs yield, +4–6 times in VFA production rate	[65]
Textile dying sludge	Ultrasonic-Fenton	Ultrasonic density= 0.14W/ml pH < 3.0	The floc structures disruption, increased from 1.48 to 6.96%	NR	NR	[66]
Concentrated sludge (40 g/L)	High-pressure homogenization (HPH)	Pressure = 150 bar, Flow rate = 2.7 m^3/h	NR	AD, full-scale, 36–38°C	+30% biogas production, +23% sludge reduction	[67]
Dewatered activated sludge (16%TS)	Thermal hydrolysis	Pilot-scale CAMBI™ Temperature = 65°C Pressure = 6 bar Duration = 20 min	Increase of VS removal from 26% to 42% (+62%)	AD, pilot-scale, treated sludge: primary sludge (80%:20%),3°C, SRT 20d	+2.3 times increase in SLR, +30–40%biogas production, improved dewaterability	[68]
Secondary sludge (30.00 g TS/L)	Thermal hydrolysis	Temperature = 134–140°C Pressure = 3.4 bar Duration = 30 min	NR	AD, batch, 35°C, HRT 30 d	+40.2% methane production, +12.6% VS removal, +6.8% digestated reduction	[69]
Secondary sludge (31.4 g TS/L)	Ultrasonication	Frequency = 20 kHz Power = 750 W Energy = 5742 kJ/kg TS	Increase of SCOD/TCOD from 0.02 to 0.10	AD, batch, 35°C, 30 d	+16.9% VS removal, +7.89 × 10−6 kWh/g energy output, 1.0 energy ratio	[69]

Type of sludge	Pre-treatment method			Biological coupling		Ref.
	Method	Condition	Outcome	Method	Outcome	
Dairy activated sludge (11.66 g TS/L	MW	Frequency = 2450 MHz Power = 900W Power = 12 min Energy = 1814 kJ/L	Increase of SCOD: 19%	AD, semi-continuous, 37°C, SRT 15 d, 170 d	+57% biogas production, +64% VS removal	[70]
WAS (35.5 ± 0.7 g TS/L)	Ultrasonication	Energy = 3380 kJ/kg TS	DDcod: 21%	AD TPAD-BMP assay (55°C→35°C)	+42% methane production, +13% VS removal	[71]
Mixed sludge	Electro kinetic disintegration	NR	Increase of SCOD: 160%, increase of DOC: 120%	AD, full-scale WWTP	+40% biogas production, biosolids requiring disposal reduced by 30%	[72]
WAS	Low thermal	Temp = 70 and 90°C Duration = 180 min	Sludge disintegration the rate was increased up to 25%	Anaerobic batch digestion	Increase in methane production = 21%	[45]
WAS	Mechanical	Mixing = 5000 rpm Duration = 10 min	Sludge disintegration the rate was increased up to 1.5 %	NR	NR	[45]
WAS	Ultrasonic-acid	Ultrasonic density= 10W/mL Duration = 10 min pH = 2.0	Sludge disintegration increased up to 40%	NR	NR	[73]
WAS	Electro kinetic disintegration	Energy = 10kWh/m^3	Increase of SCOD/TCOD to 10%, increase of SCOD from 20 to & $2gt;1000 mg/L	AD, batch, 25–30 d	+100% methane production	[74]
Sewage sludge	Thermal	Temp = 120°C Pressure = 2 atm Duration = 15 min	Soluble carbon concentration, nitrogen, and phosphorus increased by 165, 16, and 24%,	AD	Increase in methane production = 29%	[75]

Type of sludge	Pre-treatment method			Biological coupling		Ref.
	Method	Condition	Outcome	Method	Outcome	
WAS	MW—Alkali (NaOH)	Power = 900 W Temp = 95°C pH = 12	Sludge solubilization increased from 0.5 (raw) to 52.5%	Mesophilic aerobic digestion	COD degradation =81.1% VSS degradation =72.4% VSS RT = 20 d	[76]
Mixed sludge	High-pressure homogenization	Pressure = 12,000 psi Catalyst = 0.009 g NaOH/g TS	SCOD/TCOD: & 2 gt; 4.0	AD, 2TPAD, SRT 14 d, OLR 1.24±0.05 g VS/L d	0.61–1.32 L CH4/L d methane production, 43–64% VS removal, pathogen removal, net energy output	[77]
WAS	Free nitrous acid-heat pre-treatment	Nitrous acid = 0.52–1.11 mg N/L Temp = 70°C	sCOD increased and found between 0.16 and 0.28 mg sCOD/mg VS	AD	Methane production increased by = 17–26%	[78]
Dewatered sludge (16.7 ± 0.5% TS	Thermal hydrolysis	Temperature = 140–160°C Duration = 60– 90 min	Increase of DDcod from 4.5 to 34.7–42.5% (+6.7–8.4 times)	AD, batch, 37°C,28d	+~16.5% biogas production, reduction of SRT from 18–20 d to 12–14 d	[79]
Primary sludge	MW- Ultrasound	Power = 800 W Frequency = 2450 MHz, Duration = 3min US density = 0.4 W/ml US intensity = 150 W Duration = 6 min	Increase in disintegration of flocs and extracellular polymeric substances	AD	Methane production = 11.9 ml/g tCOD	[80]
Mix aerobic thicken sewage sludge	NaOH Ultrasonic	NaOH dosage = 100 g/kg Duration = 30 min Ultrasonic energy = 7500 kJ\Kg	Soluble COD increased from 275 to 6797 mg/L	Aerobic digestion	Increase in organic matter degradation = 50.7%	[81]
Sewage sludge (23 g TS/L)	High-pressure homogenization (HPH)	Pressure = 50MPa Cycles = 2	SCOD: 2167 mg/L, DDcod: 7.7%	AD, batch, 35°C, 7 d	+115% biogas production, +41.17% VS removal, +61.89% TCOD removal	[82]

Type of sludge	Pre-treatment method			Biological coupling		Ref.
	Method	Condition	Outcome	Method	Outcome	
Chemical approach						
Maize canning sludge	Ozonation	Ozone intensity = ~0.18 g O_3/g D	Increase of BOD5/COD from 26 to 93% (+2.58 times)	AD, batch,30°C, 30 d	Increase of biogas production from 1.037 to 9.52 cm^3/g COD d(+8.2 times)	[83]
WAS	Ozonation	Ozone intensity = 10 mg O_3/g TSS 20 cycles Duration = 30 s/cycle	DDcod:18%, VSS reduction:18%	AD, batch, F/I0.8, 35°C, 20 d	+800% specific biogas production, +1.6 folds VSS reduction	[84]
Activated sludge (5%TS)	Acidic treatment	8.75 mLHCl/kg wet sludge pH = 2	Four and Six times increase of soluble carbohydrate sand proteins, respectively	AD, semi-continuous, 35°C, HRT 12 d	+14.3% methane yield, –40% polymer dose for dewatering	[85]
Activated sludge	Ozonation	Ozone intensity = 0.09g O3/g MLSS, pH = 11	COD solubilization: 40%, TS reduction: 30%	AD, lab-scale AS-MBR, 120 d	Solids degradation: 37%	[86]
Sewage sludge	Alkaline treatment	0.1 mol NaOH/L	Increase of DDcod from 22.3 to 26.9%	AD, batch (BMP), 21 d	+26.4% organic removal, +1.5% biogas yield; delay of AD start up due to residual NaOH	[87]
Pulp and paper sludge	Alkaline treatment	8 g NaOH/100 g TS	Increase of SCOD:83%, 56–192% higher Sv	AD, batch, 37°C, 42 d	1040 mg acetate/L, +83% methane yield (0.32 $m^3 CH_4$/kg S removed); sodium toxicity at 16 g NaOH/100 g TS	[88]
Secondary waste water sludge	Peroxide/oxidation	60 g H_2O_2/kg TS, 0.07 g Fe2+/g H_2O_2 pH = 3	Reduction of SS:21%, reduction of VSS:25%, increase of SCOD from 0.82 to 7.8 g/ L	AD, lab-scale, 35°C, 30 d	Increase of methane production from 430 to 496 m^3 CH_4/Mg VS degraded, +3.1 time increased net energy, reduced GHG emissions (0.128 Mg CO_2/Mg of TDS)	[89]
WAS	Alkali	NaOH dosage = 157 g/ kg TS	Pre-treatments reduced the viscosity of the sludge	AD	Increase in methane production = 34%	[90]

Type of sludge	Pre-treatment method			Biological coupling		Ref.
	Method	**Condition**	**Outcome**	**Method**	**Outcome**	
Activated sludge (10.2 mg TS/L)	Peroxide/oxidation	4 g Fe3+/°/kg TS, 40 g H_2O_2/kg TS pH = 3 1 h	DDcod:23.6%, (Fe2+), DDcod:16.7% (Fe°)	AD, BMP, 35°C, 60 d	+30.2%biogas and +38.0% methane production for Fe2+, +24.4% biogas and +26.8% methane production for FeO	[91]
WAS	Fenton	Catalyst iron dosage = 4 g/kg TS H_2O_2 dosage = 40 g/kgTS pH = 3 Duration = 60 min	Sludge disintegration increased up to 23.6%	AD	Total methane production increased = 26.9%	[73]
Activated sludge (10.6 ± 0.1 g TS/L)	Alkaline treatment	pH = 9–11(4 mol/L NaOH) Duration = 24 h	NR	AD, batch,37.0 ± 0.1°C, 25 d	+10.7–13.1% TSS removal, +6.5– 12.8% VSS removal, +7.2–15.4% biogas yield, improved dewaterability	[92]
Anaerobically digested sludge	Acidic treatment	Temperature = 170°C pH = 5–6 (H_2SO_4) Duration = 1 h	NR	AD, continuous, 35°C, HRT 20 d	+2–2.5 times VSS removal, +14–21% methane production, 22–23% better dewaterability	[93]
Activated sludge	Alkaline treatment	Temperature = 130°C pH = 10 (KOH)	DDcod: around 60%	AD, continuous, 35°C, HRT 20 d	+36.4% COD removal, +33% TS removal, +74% biogas production	[94]
WAS	Free nitrous acid-heat pre-treatment	Nitrous acid = 0.52–1.11 mg N/L Temperature = 70 °C	sCOD increased and found between 0.16 and 0.28 mg sCOD/mg VS	AD	Methane production increased by = 17–26%	[78]
Sewage sludge	Ozonation	Ozone intensity = 0.1 g O_3/g COD	Oxidization of organics: 38%, solubilization of organics: 29	AD, batch, 33°C, 30 d	+1.8 times methane yield, +2.2 times production rate, decreased dewaterability	[95]
Activated sludge (11.7 ± 2.3 g TS/L)	Alkaline treatment	8 g NaOH/m_3 wet sludge (pH 8)	SCOD/TCOD: 1.99%	AD, CSTR, 55°C, HRT 21 d	+9.7% TS removal, +11.5% VS removal, +18.1% COD removal, 84.22–78.24 mL/d for biogas (−7.1%)	[96]

Type of sludge	Pre-treatment method			Biological coupling		Ref.
	Method	Condition	Outcome	Method	Outcome	
Activated sludge (13.9 ± 0.2 g TS/L)	Peroxide/oxidation	50 mg H_2O_2/g TS, 7 mg Fe/g T Sin sludge pH = 2.0 Duration = 30 min	Increase of SCOD from 8 ± 1 in control to 103 ± 7 mg/g TS (+11.9 times)	AD, BMP, 37 ± 1°C, 23 d	+10% methane production, +13% methane potential but no significant effect on hydrolysis rate	[78]
Biological approach						
WAS	Bacterial enzymaticpre-treatment	Strains = *Bacillus jerish* EDTA dosage = 0.2 g/g SS	Extracellular polymeric substance decrease to 40 mg/L	AD	Suspended solids reduction = 48.5% COD solubilization = 47.3%	[97]
Thicken sewage sludge	Bioleaching-Fenton	Bioleaching = 2 days H_2O_2 dosage= 0.12 mol/L Fe2+ dosage = 0.036 mol/L Duration = 60 min	Volatile solids reduction up to 36.93%. Sludge resistance to filtration was 3.43 × 108 s2/g. Increased dewater ability by 4%	NR	NR	[98]

NR, not reported; AD, anaerobic digestion; RT, retention time; MW, microwave; WAS, waste activated sludge.

Table 1.
Methods for activated sludge treatment.

reaction of hydroxyl radicals with organic compounds present in WAS. Hydroxyl radicals are highly reactive species and may cause complete mineralization of WAS after oxidation [101]. Ozone forms free radicals on reacting with water and causes hydrolysis of organic matter in WAS enhancing its biodegradability [102]. Chemical pre-treatment is not suitable for easily biodegradable substances containing high amounts of carbohydrates, due to their accelerated degradation and subsequent accumulation of volatile fatty acids, which leads to failure of the methanogenesis [103].

2.3.2 Acid pre-treatment

Acid pre-treatment is done to disintegrate the polymeric structures and cells in WAS which is achieved by the use of reagents such as HCl, H_2SO_4, H_3PO_4, and HNO_2. The pH during the acid pre-treatment ranges from 1 to 5.5. During acid pre-treatment, flocculation is observed near isoelectric point as the lowering of pH causes reaction between hydrogen ions and the ionized carboxyl groups rendering them in unionized forms resulting in the formation of aggregates [104]. Strong acidic pre-treatment may result in the production of inhibitory by-products, such as furfural and hydroxyl-methylfurfural [105]. Hence, strong acidic pre-treatment is avoided and pre-treatment with dilute acids is coupled with thermal methods. Other disadvantages associated with acid pre-treatment include the loss of fermentable sugar due to the increased degradation of complex substrates, a high cost of acids, and the additional cost for neutralizing the acidic conditions before the AD process [106].

2.3.3 Alkali pre-treatment

Alkali treatment is relatively effective in sludge solubilization, within the order of efficacy being highest for NaOH followed by KOH, $Mg(OH)_2$ and $Ca(OH)_2$ [107]. However, too high concentrations of Na^+ or K^+ may cause subsequent inhibition of AD [107]. The increase in pH of WAS due to alkali pre-treatment causes many effects on sludge components which include saponification of lipid bilayer and protein denaturation in the cell membrane, solubilizing EPS by the ionization of its carboxyl and amino groups and hydrolysis of sludge organic substances [99].

An alkali pre-treatment study demonstrated that the best-performing alkali for WAS is NaOH. The results of this study indicated an increase by 39.8, 36.6, 15.3, and 10.8% of the soluble COD (mg/L) for WAS by using NaOH, KOH, $Ca(OH)_2$, and $Mg(OH)_2$, respectively [108]. Using 8% of NaOH, an increase in the methane yield by 81% was observed for pulp and paper sludge [88]. Moreover, these pre-treatment methods were further studied with the pH range between 4 and 11 [109]. The results indicated that acidic pre-treatment was less effective than the alkali pre-treatment method for soluble COD in short-chain fatty acids from excess sludge. The main disadvantage of this pre-treatment includes additional pH adjustment need of this pre-treatment for AD which increases operational cost and also increases environmental concerns due to additional chemical agents.

2.3.4 Oxidation

The COD removal during AD was enhanced through oxidation at 90°C with 2 gH_2O_2/g VSS (volatile suspended solids) but not by the oxidation at 37°C [110]. Moreover, post-treatment on the recirculation loop, treating 20% of the sludge stream, was more efficient than a configuration with pre-treatment. However, the process consisting of one anaerobic digester, high-temperature oxidation and a second digester led to the highest removal of fecal coliforms [110]. Fenton reaction involves the decomposition of hydrogen peroxide in the presence of ferrous ions as the catalyst to form hydroxyl radicals [111]. The hydroxyl radicals thus formed are highly reactive free radical species that oxidize organic matter in sludge further enhancing WAS biodegradability and dewatering [112]. Besides catalyst and hydrogen peroxide, pH during the reaction is also a very crucial parameter to be maintained during Fenton oxidation as the catalytic activity of ferrous ions is lost at pH > 4 [113]. Hence, an effective Fenton oxidation involves adjustment to acidic pH values, oxidation, neutralization, and separation of by-products [114]. Use of Fenton catalyzed oxidation (0.067 g Fe(II)/g H_2O_2, and 60 g H_2O_2/kg TS) decreased sludge resistance to dewatering in terms of capillary suction time, but did not have a positive effect on sludge dewatering performance on a belt press simulation [115].

2.3.5 Ozonation

Ozonation depends upon the oxidation reaction of hydroxyl radicals with organic compounds present in WAS. Ozone forms free radicals on reacting with water and causes hydrolysis of organic matter in WAS enhancing its biodegradability [102]. The pH of the system is reduced after ozonation because ozone degrades higher molecular weight organic compounds into simpler acidic compounds like carboxylic acid [11]. Ozone is a strong oxidant, which disintegrates itself into radicals and reacts with organic substrates [116] in two ways; the direct reaction depends on the structure of the reactant, whereas the indirect reaction is based on the hydroxyl radicals. As a result, the recalcitrant compounds become more biodegradable and accessible to the anaerobic bacteria [117]. Prior to ozone treatment, the methane production was observed to be 440.3 ml CH_4/g VS and after applying ozone doses of 0.034 g O_3/g TS, 0.068 g O_3/g TS, 0.101 g O_3/g TS, and 0.202 g O_3/g TS increased by 35.2, 46.4, 32.9, and 22.2%, respectively [118]. Several ozonation pre-treatment systems are commercially available in the market. They include the Aspal SLUDGE™ and Praxair® Lyso™. The former offers high dewaterability and low energy consumption, and the latter achieves 80% sludge reduction and 75% reduction in ozone use with increasing dewaterability of sludge [112].

2.3.6 Temperature phased AD (TPAD)

Temperature phased AD (TPAD) occurs in two phases. In the primary hydrolytic/acidogenic phase, 45–70°C temperatures for 2–6 d is applied whereas the second phase is the methanogenic or acetogenic phase for which temperature favorable to thermophilic microorganisms is provided for 14–30 d. The effects of Maillard reaction have not been reported during TPAD which may be due to increased activity of hydrolytic enzymes or defense created by microorganisms through enzymes suppressing the effects of Maillard reaction products [119]. An additional acidification step decreases the amount of polyelectrolyte required to dewater the digestate since poor dewaterability is observed in the acidogenic effluent.

3. Other different pre-treatment methods

3.1 Thermochemical pre-treatment

Integration of pre-treatment methods has also been studied for sludge stabilization to further improve AD and biogas production. A combination of thermal and chemical pre-treatment methods is known to improve the degradation of volatile solids and biogas production [108]. Thermal-alkaline pre-treatment of sludge was reported to cause floc disintegration, cell disruption and reduction in organic sludge components with high increase in sludge pH to 13 [120]. The authors also reported about 100 times increase in SCOD of sludge as compared to raw untreated sludge [120]. The improvement in reduction of volatile solids as a result of thermochemical pre-treatment enhanced two times high reduction in volatile solids than that in control when sodium hydroxide was combined with thermal pre-treatment at 121°C [108]. In a similar study, 72% enhancement in volatile solids removal and biogas production was observed when sludge was pre-treated at 170°C and pH 12 [94]. The improved content of soluble COD after thermochemical pre-treatment plays role in increasing the efficiency of AD with biogas production increased to 52.78% [2]. Chemical pre-treatment of carbohydrates and proteins can increase their hydrolysis into sugars and amino acids, respectively, and these later products react with each other through Maillard reaction at high temperature resulting in high molecular weight polymers like melanoidins. In another study, a high 78% biogas production with 60% methane was obtained after thermochemical pre-treatment at a lower temperature of 70°C [121].

Microwave-alkaline pre-treatment is another integrated technology for sludge pre-treatment which improves the efficiency of AD. Microwave irradiation coupled with alkaline pre-treatment of sludge improved the volatile solids reduction by 35% and methane formation by 53% as compared to control [122].

4. Feasibility of a full-scale application

With an ever-increasing concern for the environment, different pre-treatment methods can enhance the AD performance. Nevertheless, the high capital cost, high consumption of energy, required chemicals, and sophisticated operating conditions (maintenance, odor control, etc.) are the major factor hindering their full-scale application [123]. There are only a few examples of the thermal hydrolysis process (THP) that have been applied at full-scale such as the Cambi, Porteous, and Zimpro process and thermochemical pre-treatment methods such as Synox, Protox, and Krepro. It should be noted that these methods are all applied for WWTP sludge. Concerning the organic fraction of municipal solid wastes, only a few mechanical pre-treatment methods such as Cambi THP and AD with a pre-hydrolysis stage (two-stage AD) have been applied at a full scale.

4.1 Energy balance

The required energy depends on the desired pre-treatment temperature. If it is above 100°C, most of the energy is utilized in water vaporization, thus making it less desirable [124]. Microwave heating provides direct heating from the inside and therefore unlike conventional heating strategies, negligible or no heat losses are reported [125]. However, neither microwave nor ultrasound were found to be energy-intensive for pretreating mixed sludge, as the enhanced methane yields were not enough to compensate for the required energy [126]. The total biogas obtained after thermal pre-treatment of sludge is relatively higher than other methods and thus costs could be compensated by utilizing the extra biogas through an efficient heat exchanger [127]. A better energy balance was estimated while treatment of organic fraction of municipal solid waste in two-stage AD systems where the authors observed higher energy potential to be associated with not the first stage hydrogen production but to the higher performance in the methanogenic reactor [128]. The energy efficiency of a two-stage AD reactor for sewage sludge in which excess energy of 2.17 kJ/d was obtained was higher when compared to a single-stage system [129]. It was further concluded that the energy balance can be enhanced by 18.5% if the two-stage AD process is optimized.

4.2 Economic feasibility

Estimated net profit of various pre-treatments (low-temperature thermal pre-treatment not included) to enhance the biogas production of food waste obtained the best result (10–15 euro/ton FW) with less energy-intensive methods (acid and freeze-thaw) [130]. The estimation of the economic feasibility of pre-treatment methods based on a full-scale application has only been reported for WWTP sludge. The operational and maintenance cost of a full-scale AD (3300 m^3) treating 380 m^3 sludge per day based on the application of focused-pulsed pre-treatment technology could generate a benefit of 540,000 USD per year [72]. An approximate cost estimate associated with pre-treatment methods was suggested in research which included capital, operational, and maintenance costs between 70 and 150 US $/ton sludge [131]. In another cost estimate study, comparative costs for sludge pre-treatment methods were calculated to improve the process of AD of sludge where the authors estimated costs associated with microwave pre-treatment, conventional thermal pre-treatment, ultrasound, and chemical pre-treatment methods as 0.0162, 0.0187, 0.0264, and 0.0358 US$/$m^3$, respectively [125]. The comparative cost analysis suggested microwave and conventional pre-treatment methods to be cheaper than ultrasonic and chemical pre-treatments.

The amount of sludge for pre-treatment is also an important factor to consider when estimating the pre-treatment cost. Pre-treatment strategies such as ultrasound can prove to be energetically acceptable for a large scale application if 6 kWh energy value is considered for each cubic meter of sludge [124]. If higher energy is required, biological pre-treatment such as adding hydrolytic bacteria could be a cheaper option [132]. The extent of net economic benefit also depends upon other factors besides the type of pre-treatment method and quality and quantity of sludge. Other parameters to be considered include treatment capacity, availability of labor, the cost associated with collection and transport, taxes and tariffs, energy prices, price of land selected for setup, costs for additional mixing and pumping requirements, the market value of end product as well as waste and residue disposal [133].

5. Conclusion

The pre-treatment methods have the potential to solubilize complex sludge components which include the organic matter, EPS, and the microbial cell wall which in turn makes the progression of subsequent biological degradation treatments easier. AD is the sustainable process widely employed for bioenergy generation from the WAS. Further pre-treatment enhances the methane percentage in the biogas thus the process is energy efficient and sustainable. To attain a clear understanding of the mechanism behind each method, the focus should be given to the conversion strategy and structural alterations that occur in complex WAS components upon the application of each pretreatment technology. Many physical, chemical and biological pretreatment methods have been mentioned in the literature that has been used individually but each of them owns certain disadvantages. These limitations range from high energy requirement of microwaves to excess degradation and fermentable sugars loss in acid pre-treatment. Using two pre-treatment methods in combination has known to overcome these problems, reaping the efficiency of both methods simultaneously. Thus combing pre-treatment process and AD will lead to sustainable process for sludge management.

Author details

Sridhar Pilli[1]*, Ashutosh Kumar Pandey[2], Ankur Katiyar[3], Kritika Pandey[4] and Rajeshwar Dayal Tyagi[5]

1 Department of Civil Engineering, National Institute of Technology, Warangal, Warangal, India

2 Centre for Energy and Environmental Sustainability, Lucknow, India

3 Department of Biotechnology, S.D. College of Engineering and Technology, Muzaffarnagar, Uttar Pradesh, India

4 Department of Biotechnology, Dr. Ambedkar Institute of Technology for Handicapped, Kanpur, Uttar Pradesh, India

5 Institut National de la Recherche Scientifique, Quebec City, Quebec, Canada

*Address all correspondence to: srenitw@nitw.ac.in

References

[1] Liu G et al. Scenarios for sewage sludge reduction and reuse in clinker production towards regional eco-industrial development: A comparative emergy-based assessment. Journal of Cleaner Production. 2015;**103**:371-383. DOI: 10.1016/j.jclepro.2014.09.003

[2] Xu J, Yuan H, Lin J, Yuan W. Evaluation of thermal, thermal-alkaline, alkaline and electrochemical pretreatments on sludge to enhance anaerobic biogas production. Journal of the Taiwan Institute of Chemical Engineers. 2014;**45**(5):2531-2536. DOI: 10.1016/j.jtice.2014.05.029

[3] Xu Y, Lu Y, Zheng L, Wang Z, Dai X. Perspective on enhancing the anaerobic digestion of waste activated sludge. Journal of Hazardous Materials. 2020;**389**:121847. DOI: 10.1016/j.jhazmat.2019.121847

[4] Hanjie Z. Sludge treatment to increase biogas production. Trita-LWR Degree Project 10-20. In: Skolan för Arkitektur Och Samhällsbyggnad. Kungliga Tekniska Högskolan. Stockholm, Sweden: Department of Land and Water Resources Engineering: Royal Institute of Technology (KTH) SE-100 44; 2010

[5] Elliott A, Mahmood T. Comparison of mechanical pretreatment methods for the enhancement of anaerobic digestion of pulp and paper waste activated sludge. Water Environment Research. 2012;**84**(6):497-505. DOI: 10.2175/106143012x13347678384602

[6] Esposito G, Frunzo L, Panico A, Pirozzi F. Modelling the effect of the OLR and OFMSW particle size on the performances of an anaerobic co-digestion reactor. Process Biochemistry. 2011;**46**(2):557-565. DOI: 10.1016/j.procbio.2010.10.010

[7] Kim IS, Kim DH, Hyun SH. Effect of particle size and sodium ion concentration on anaerobic thermophilic food waste digestion. Water Science and Technology. 2000;**41**(3):67-73. DOI: 10.2166/wst.2000.0057

[8] Khanal SK, Grewell D, Sung S, Van Leeuwen J. Ultrasound applications in wastewater sludge pretreatment: A review. Critical Reviews in Environmental Science and Technology. 2007;**37**(4):277-313. DOI: 10.1080/10643380600860249

[9] Delmas H, Le NT, Barthe L, Julcour-Lebigue C. Optimization of hydrostatic pressure at varied sonication conditions–power density, intensity, very low frequency–for isothermal ultrasonic sludge treatment. Ultrasonics Sonochemistry. 2015;**25**:51-59

[10] Chu CP, Lee DJ, Chang BV, You CS, Tay JH. 'Weak' ultrasonic pre-treatment on anaerobic digestion of flocculated activated biosolids. Water Research. 2002;**36**(11):2681-2688. DOI: 10.1016/S0043-1354(01)00515-2

[11] Bougrier C, Albasi C, Delgenès JP, Carrère H. Effect of ultrasonic, thermal and ozone pre-treatments on waste activated sludge solubilisation and anaerobic biodegradability. Chemical Engineering and Processing Process Intensification. 2006;**45**(8):711-718. DOI: 10.1016/j.cep.2006.02.005

[12] Kumar PS, Pandit AB. Modeling hydrodynamic cavitation. Chemical Engineering and Technology: Industrial Chemistry Equipment-Process Engineering. 1999;**22**(12):1017-1027

[13] Zhang Y, Zhang P, Ma B, Wu H, Zhang S, Xu X. Sewage sludge disintegration by high-pressure homogenization: A sludge disintegration model. Journal of Environmental Sciences. 2012;**24**(5):814-820. DOI: 10.1016/S1001-0742(11)60834-6

[14] Barjenbruch M, Kopplow O. Enzymatic, mechanical and thermal

pre-treatment of surplus sludge. Advances in Environmental Research. 2003;**7**(3):715-720. DOI: 10.1016/S1093-0191(02)00032-1

[15] Carlsson M, Lagerkvist A, Ecke H, editors. Electroporation for enhanced methane yield from municipal solid waste. In: ORBIT 2008: Moving Organic Waste Recycling Towards Resource Management and Biobased Economy 13/10/2008-15/10/2008. Sweden: University of Technology; 2008

[16] Engelhart M, Krüger M, Kopp J, Dichtl N. Effects of disintegration on anaerobic degradation of sewage excess sludge in downflow stationary fixed film digesters. Water Science and Technology. 2000;**41**(3):171-179. DOI: 10.2166/wst.2000.0069

[17] Toreci I, Kennedy KJ, Droste RL. Evaluation of continuous mesophilic anaerobic sludge digestion after high temperature microwave pretreatment. Water Research. 2009;**43**(5):1273-1284. DOI: 10.1016/j.watres.2008.12.022

[18] Attard P, Moody MP, Tyrrell JWG. Nanobubbles: The big picture. Physica A: Statistical Mechanics and its Applications. 2002;**314**(1-4):696-705

[19] Zhu J et al. Cleaning with bulk nanobubbles. Langmuir. 2016;**32**(43):11203-11211

[20] Vaze N et al. An integrated electrolysis – electrospray – ionization antimicrobial platform using Engineered Water Nanostructures (EWNS) for food safety applications. Food Control. 2018;**85**:151-160. DOI: 10.1016/j.foodcont.2017.09.034

[21] Agarwal A, Ng WJ, Liu Y. Principle and applications of microbubble and nanobubble technology for water treatment. Chemosphere. 2011;**84**(9):1175-1180. DOI: 10.1016/j.chemosphere.2011.05.054

[22] Oeffinger BE, Wheatley MA. Development and characterization of a nano-scale contrast agent. Ultrasonics. 2004;**42**(1-9):343-347. DOI: 10.1016/j.ultras.2003.11.011

[23] Kim JY, Song MG, Kim JD. Zeta potential of nanobubbles generated by ultrasonication in aqueous alkyl polyglycoside solutions. Journal of Colloid and Interface Science. 2000;**223**(2):285-291. DOI: 10.1006/jcis.1999.6663

[24] Liu S, Oshita S, Kawabata S, Makino Y, Yoshimoto T. Identification of ROS produced by nanobubbles and their positive and negative effects on vegetable seed germination. Langmuir. 2016;**32**(43):11295-11302. DOI: 10.1021/acs.langmuir.6b01621

[25] Yang X et al. Enhanced hydrolysis of waste activated sludge for methane production via anaerobic digestion under N2-nanobubble water addition. Science of the Total Environment. 2019;**693**:133524. DOI: 10.1016/j.scitotenv.2019.07.330

[26] Brennen CE. Cavitation and Bubble Dynamics. New York, USA: Cambridge University Press; 2014

[27] Ozonek J. Application of Hydrodynamic Cavitation in Environmental Engineering. Boca Raton, FL: CRC Press; 2012

[28] Lee I, Han JI. The effects of waste-activated sludge pretreatment using hydrodynamic cavitation for methane production. Ultrasonics Sonochemistry. 2013;**20**(6):1450-1455. DOI: 10.1016/j.ultsonch.2013.03.006

[29] Kim HJ, Nguyen DX, Bae JH. The performance of the sludge pretreatment system with venturi tubes. Water Science and Technology. 2008;**57**(1):131-137. DOI: 10.2166/wst.2008.717

[30] Petkovšek M, Mlakar M, Levstek M, Stražar M, Širok B, Dular M. A novel rotation generator of hydrodynamic

cavitation for waste-activated sludge disintegration. Ultrasonics Sonochemistry. 2015;**26**:408-414. DOI: 10.1016/j.ultsonch.2015.01.006

[31] Neumann P, Pesante S, Venegas M, Vidal G. Developments in pre-treatment methods to improve anaerobic digestion of sewage sludge. Reviews in Environmental Science and Biotechnology. 2016;**15**(2):173-211. DOI: 10.1007/s11157-016-9396-8

[32] Souza TSO, Ferreira LC, Sapkaite I, Pérez-Elvira SI, Fdz-Polanco F. Thermal pretreatment and hydraulic retention time in continuous digesters fed with sewage sludge: Assessment using the ADM1. Bioresource Technology. 2013;**148**:317-324. DOI: 10.1016/j.biortech.2013.08.161

[33] Nazari L et al. Low-temperature thermal pre-treatment of municipal wastewater sludge: Process optimization and effects on solubilization and anaerobic degradation. Water Research. 2017;**113**:111-123. DOI: 10.1016/j.watres.2016.11.055

[34] Vaclavik VA, Christian EW, Christian EW. Essentials of Food Science. Vol. 42. New York: Springer Science+Business Media; 2008

[35] Koupaie EH, Eskicioglu C. Conventional heating vs. microwave sludge pretreatment comparison under identical heating/cooling profiles for thermophilic advanced anaerobic digestion. Waste Management. 2016;**53**:182-195

[36] Pilli S, Yan S, Tyagi RD, Surampalli RY. Thermal pretreatment of sewage sludge to enhance anaerobic digestion: A review. Critical Reviews in Environmental Science and Technology. 2015;**45**(6):669-702. DOI: 10.1080/10643389.2013.876527

[37] Sapkaite I, Barrado E, Fdz-Polanco F, Pérez-Elvira SI. Optimization of a thermal hydrolysis process for sludge pre-treatment. Journal of Environmental Management. 2017;**192**:25-30. DOI: 10.1016/j.jenvman.2017.01.043

[38] Carrère H et al. Pretreatment methods to improve sludge anaerobic degradability: A review. Journal of Hazardous Materials. 2010;**183**(1-3):1-15. DOI: 10.1016/j.jhazmat.2010.06.129

[39] Wilson CA, Novak JT. Hydrolysis of macromolecular components of primary and secondary wastewater sludge by thermal hydrolytic pretreatment. Water Research. 2009;**43**(18):4489-4498

[40] Graja S, Chauzy J, Fernandes P, Patria L, Cretenot D. Reduction of sludge production from WWTP using thermal pretreatment and enhanced anaerobic methanisation. Water Science and Technology. 2005;**52**(1-2):267-273. DOI: 10.2166/wst.2005.0527

[41] Dwyer J, Starrenburg D, Tait S, Barr K, Batstone DJ, Lant P. Decreasing activated sludge thermal hydrolysis temperature reduces product colour, without decreasing degradability. Water Research. 2008;**42**(18):4699-4709. DOI: 10.1016/j.watres.2008.08.019

[42] Gianico A, Braguglia CM, Cesarini R, Mininni G. Reduced temperature hydrolysis at 134°C before thermophilic anaerobic digestion of waste activated sludge at increasing organic load. Bioresource Technology. 2013;**143**:96-103. DOI: 10.1016/j.biortech.2013.05.069

[43] Liao X, Li H, Zhang Y, Liu C, Chen Q. Accelerated high-solids anaerobic digestion of sewage sludge using low-temperature thermal pretreatment. International Biodeterioration and Biodegradation. 2016;**106**:141-149. DOI: 10.1016/j.ibiod.2015.10.023

[44] Carvajal A, Peña M, Pérez-Elvira S. Autohydrolysis pretreatment of secondary sludge for anaerobic

digestion. Biochemical Engineering Journal. 2013;**75**:21-31. DOI: 10.1016/j.bej.2013.03.002

[45] Ruffino B et al. Improvement of anaerobic digestion of sewage sludge in a wastewater treatment plant by means of mechanical and thermal pre-treatments: Performance, energy and economical assessment. Bioresource Technology. 2015;**175**:298-308. DOI: 10.1016/j.biortech.2014.10.071

[46] Hiraoka M, Takeda N, Sakai S, Yasuda A. Highly efficient anaerobic digestion with thermal pretreatment. Water Science and Technology. 1985;**17**(4-5):529-539. DOI: 10.2166/wst.1985.0157

[47] Prorot A, Julien L, Christophe D, Patrick L. Sludge disintegration during heat treatment at low temperature: A better understanding of involved mechanisms with a multiparametric approach. Biochemical Engineering Journal. 2011;**54**(3):178-184. DOI: 10.1016/j.bej.2011.02.016

[48] Skiadas IV, Gavala HN, Lu J, Ahring BK. Thermal pre-treatment of primary and secondary sludge at 70 C prior to anaerobic digestion. Water Science and Technology. 2005;**52**(1-2):161-166

[49] Appels L, Houtmeyers S, Degrève J, Van Impe J, Dewil R. Influence of microwave pre-treatment on sludge solubilization and pilot scale semi-continuous anaerobic digestion. Bioresource Technology. 2013;**128**:598-603. DOI: 10.1016/j.biortech.2012.11.007

[50] Braguglia CM, Gianico A, Mininni G. Laboratory-scale ultrasound pre-treated digestion of sludge: heat and energy balance. Bioresource Technology. 2011;**102**(16):7567-7573. DOI: 10.1016/j.biortech.2011.05.025

[51] Burger G, Parker W. Investigation of the impacts of thermal pretreatment on waste activated sludge and development of a pretreatment model. Water Research. 2013;**47**(14):5245-5256. DOI: 10.1016/j.watres.2013.06.005

[52] Chang CJ, Tyagi VK, Lo SL. Effects of microwave and alkali induced pretreatment on sludge solubilization and subsequent aerobic digestion. Bioresource Technology. 2011;**102**(17):7633-7640. DOI: 10.1016/j.biortech.2011.05.031

[53] Chang TC, You SJ, Damodar RA, Chen YY. Ultrasound pre-treatment step for performance enhancement in an aerobic sludge digestion process. Journal of the Taiwan Institute of Chemical Engineers. 2011;**42**(5):801-808. DOI: 10.1016/j.jtice.2011.01.003

[54] Choi H, Jeong SW, Chung YJ. Enhanced anaerobic gas production of waste activated sludge pretreated by pulse power technique. Bioresource Technology. 2006;**97**(2):198-203. DOI: 10.1016/j.biortech.2005.02.023

[55] Coelho NMG, Droste RL, Kennedy KJ. Evaluation of continuous mesophilic, thermophilic and temperature phased anaerobic digestion of microwaved activated sludge. Water Research. 2011;45(9):2822-2834. DOI: 10.1016/j.watres.2011.02.032

[56] Ebenezer AV, Arulazhagan P, Adish Kumar S, Yeom IT, Rajesh Banu J. Effect of deflocculation on the efficiency of low-energy microwave pretreatment and anaerobic biodegradation of waste activated sludge. Applied Energy. 2015;**145**:104-110. DOI: 10.1016/j.apenergy.2015.01.133

[57] Erden G. Combination of alkaline and microwave pretreatment for disintegration of meat processing wastewater sludge. Environmental Technology (United Kingdom). 2013;**34**(6):711-718. DOI: 10.1080/09593330.2012.715678

[58] Gong C, Jiang J, Li D. Ultrasound coupled with Fenton oxidation pre-treatment of sludge to release organic carbon, nitrogen and phosphorus. Science of the Total Environment. 2015;**532**:495-500. DOI: 10.1016/j.scitotenv.2015.05.131

[59] Houtmeyers S, Degrève J, Willems K, Dewil R, Appels L. Comparing the influence of low power ultrasonic and microwave pre-treatments on the solubilisation and semi-continuous anaerobic digestion of waste activated sludge. Bioresource Technology. 2014;**171**:44-49. DOI: 10.1016/j.biortech.2014.08.029

[60] Ki D, Parameswaran P, Popat SC, Rittmann BE, Torres CI. Effects of pre-fermentation and pulsed-electric-field treatment of primary sludge in microbial electrochemical cells. Bioresource Technology. 2015;**195**:83-88. DOI: 10.1016/j.biortech.2015.06.128

[61] Lee IS, Rittmann BE. Effect of low solids retention time and focused pulsed pre-treatment on anaerobic digestion of waste activated sludge. Bioresource Technology. 2011;**102**(3):2542-2548. DOI: 10.1016/j.biortech.2010.11.082

[62] Liu C, Lei Z, Yang Y, Zhang Z. Preliminary trial on degradation of waste activated sludge and simultaneous hydrogen production in a newly-developed solar photocatalytic reactor with AgX/TiO2-coated glass tubes. Water Research. 2013;**47**(14):4986-4992

[63] Liu C, Shi W, Li H, Lei Z, He L, Zhang Z. Improvement of methane production from waste activated sludge by on-site photocatalytic pretreatment in a photocatalytic anaerobic fermenter. Bioresource Technology. 2014;**155**:198-203. DOI: 10.1016/j.biortech.2013.12.041

[64] Martín MÁ, González I, Serrano A, Siles JÁ. Evaluation of the improvement of sonication pre-treatment in the anaerobic digestion of sewage sludge. Journal of Environmental Management. 2015;**147**:330-337. DOI: 10.1016/j.jenvman.2014.09.022

[65] Morgan-Sagastume F, Pratt S, Karlsson A, Cirne D, Lant P, Werker A. Production of Volatile Fatty Acids by Fermentation of Waste Activated Sludge Pre-Treated in Full-Scale Thermal Hydrolysis Plants. Bioresource Technology. 2011;**102**(3):3089-3097. DOI: 10.1016/j.biortech.2010.10.054

[66] Ning XA, Chen H, Wu J, Wang Y, Liu J, Lin M. Effects of ultrasound assisted fenton treatment on textile dyeing sludge structure and dewaterability. Chemical Engineering Journal. 2014;242:102-108. DOI: 10.1016/j.cej.2013.12.064

[67] Onyeche T. Economic benefits of low pressure sludge homogenization for wastewater treatment plants. In: IWA Specialist Conferences. Moving Forward Wastewater Biosolids Sustainability; Moncton, New Brunswick, Canada; June 2007

[68] Oosterhuis M, Ringoot D, Hendriks A, Roeleveld P. Thermal hydrolysis of waste activated sludge at Hengelo wastewater treatment plant. The Netherlands: Water science and technology. 2014;**70**(1):1-7

[69] Pilli S, Yan S, Tyagi RD, Surampalli RY. Anaerobic digestion of ultrasonicated sludge at different solids concentrations—computation of mass-energy balance and greenhouse gas emissions. Journal of Environmental Management. 2016;**166**:374-386. DOI: 10.1016/j.jenvman.2015.10.041

[70] Uma Rani R, Adish Kumar S, Kaliappan S, Yeom IT, Rajesh Banu J. Impacts of microwave pretreatments on the semi-continuous anaerobic digestion of dairy waste activated sludge. Waste Management. 2013;**33**(5):1119-1127. DOI: 10.1016/j.wasman.2013.01.016

[71] Wang Q, Kuninobu M, Kakimoto K, Ogawa HI, Kato Y. Upgrading of anaerobic digestion of waste activated sludge by ultrasonic pretreatment. Bioresource Technology. 1999;**68**(3):309-313. DOI: 10.1016/S0960-8524(98)00155-2

[72] Rittmann BE, Lee HS, Zhang H, Alder J, Banaszak JE, Lopez R. Full-scale application of focused-pulsed pre-treatment for improving biosolids digestion and conversion to methane. Water Science and Technology. 2008;**58**(10):1895-1901. DOI: 10.2166/wst.2008.547

[73] Sahinkaya S. Disintegration of municipal waste activated sludge by simultaneous combination of acid and ultrasonic pretreatment. Process Safety and Environment Protection. 2015;**93**:201-205. DOI: 10.1016/j.psep.2014.04.002

[74] Salerno MB, Lee H-S, Parameswaran P, Rittmann BE. Using a pulsed electric field as a pretreatment for improved biosolids digestion and methanogenesis. Water Environment Research. 2009;**81**(8):831-839. DOI: 10.2175/106143009x407366

[75] Serrano A, Siles JA, Gutiérrez MC, Martín MÁ. Improvement of the biomethanization of sewage sludge by thermal pre-treatment and co-digestion with strawberry extrudate. Journal of Cleaner Production. 2015;**90**:25-33. DOI: 10.1016/j.jclepro.2014.11.039

[76] Tyagi VK, Lo SL. Enhancement in mesophilic aerobic digestion of waste activated sludge by chemically assisted thermal pretreatment method. Bioresource Technology. 2012;**119**:105-113. DOI: 10.1016/j.biortech.2012.05.134

[77] Wahidunnabi AK, Eskicioglu C. High pressure homogenization and two-phased anaerobic digestion for enhanced biogas conversion from municipal waste sludge. Water Research. 2014;**66**:430-446. DOI: 10.1016/j.watres.2014.08.045

[78] Wang Q, Jiang G, Ye L, Yuan Z. Enhancing methane production from waste activated sludge using combined free nitrous acid and heat pre-treatment. Water Research. 2014;**63**:71-80. DOI: 10.1016/j.watres.2014.06.010

[79] Xue Y, Liu H, Chen S, Dichtl N, Dai X, Li N. Effects of thermal hydrolysis on organic matter solubilization and anaerobic digestion of high solid sludge. Chemical Engineering Journal. 2015;**264**:174-180. DOI: 10.1016/j.cej.2014.11.005

[80] Mesfin Yeneneh A, Kanti Sen T, Chong S, Ming Ang H, Kayaalp A. Effect of combined microwave-ultrasonic pretreatment on anaerobic biodegradability of primary, excess activated and mixed sludge. Computational Water, Energy, and Environmental Engineering. 2013;**2**:7-11. DOI: 10.4236/cweee.2013.23b002

[81] Jin Y, Li H, Mahar RB, Wang Z, Nie Y. Combined alkaline and ultrasonic pretreatment of sludge before aerobic digestion. Journal of Environmental Sciences. 2009;**21**(3):279-284. DOI: 10.1016/S1001-0742(08)62264-0

[82] Zhang S, Zhang P, Zhang G, Fan J, Zhang Y. Enhancement of anaerobic sludge digestion by high-pressure homogenization. Bioresource Technology. 2012;**118**:496-501. DOI: 10.1016/j.biortech.2012.05.089

[83] Beszédes S, Kertész S, László Z, Szabo G, Hodur C. Biogas production of ozone and/or microwave-pretreated canned maize production sludge. Ozone Science and Engineering. 2009;**31**(3):257-261

[84] Cheng CJ, Hong PKA. Anaerobic digestion of activated sludge after pressure-assisted ozonation. Bioresource Technology. 2013;**142**:69-76. DOI: 10.1016/j.biortech.2013.04.058

[85] Devlin DC, Esteves SRR, Dinsdale RM, Guwy AJ. The effect of

acid pretreatment on the anaerobic digestion and dewatering of waste activated sludge. Bioresource Technology. 2011;**102**(5):4076-4082. DOI: 10.1016/j.biortech.2010.12.043

[86] Kumar MSK, Kumar TK, Arulazhagan P, Kumar SA, Yeom IT, Banu JR. Effect of alkaline and ozone pretreatment on sludge reduction potential of a membrane bioreactor treating high-strength domestic wastewater. Desalination and Water Treatment. 2015;**55**(5):1127-1134. DOI: 10.1080/19443994.2014.923335

[87] Li H, Li C, Liu W, Zou S. Optimized alkaline pretreatment of sludge before anaerobic digestion. Bioresource Technology. 2012;**123**:189-194. DOI: 10.1016/j.biortech.2012.08.017

[88] Lin Y, Wang D, Wu S, Wang C. Alkali pretreatment enhances biogas production in the anaerobic digestion of pulp and paper sludge. Journal of Hazardous Materials. 2009;**170**(1):366-373. DOI: 10.1016/j.jhazmat.2009.04.086

[89] Pilli S, More TT, Yan S, Tyagi RD, Surampalli RY. Fenton pre-treatment of secondary sludge to enhance anaerobic digestion: Energy balance and greenhouse gas emissions. Chemical Engineering Journal. 2016;**283**:285-292. DOI: 10.1016/j.cej.2015.07.056

[90] Ruiz-Hernando M et al. Effect of ultrasound, low-temperature thermal and alkali pre-treatments on waste activated sludge rheology, hygienization and methane potential. Water Research. 2014;**61**:119-129. DOI: 10.1016/j.watres.2014.05.012

[91] Şahinkaya S, Kalipci E, Aras S. Disintegration of waste activated sludge by different applications of fenton process. Process Safety and Environment Protection. 2015;**93**:274-281. DOI: 10.1016/j.psep.2014.05.010

[92] Shao L, Wang X, Xu H, He P. Enhanced anaerobic digestion and sludge dewaterability by alkaline pretreatment and its mechanism. Journal of Environmental Sciences (China). 2012;**24**(10):1731-1738. DOI: 10.1016/S1001-0742(11)61031-0

[93] Takashima M, Tanaka Y. Acidic thermal post-treatment for enhancing anaerobic digestion of sewage sludge. Journal of Environmental Chemical Engineering. 2014;**2**(2):773-779. DOI: 10.1016/j.jece.2014.02.018

[94] Valo A, Carrère H, Delgenès JP. Thermal, chemical and thermo-chemical pre-treatment of waste activated sludge for anaerobic digestion. Journal of Chemical Technology and Biotechnology. 2004;**79**(11):1197-1203. DOI: 10.1002/jctb.1106

[95] Weemaes M, Grootaerd H, Simoens F, Verstraete W. Anaerobic digestion of ozonized biosolids. Water Research. 2000;**34**(8):2330-2336. DOI: 10.1016/S0043-1354(99)00373-5

[96] Wonglertarak W, Wichitsathian B. Alkaline pretreatment of waste activated sludge in anaerobic digestion. Journal of Clean Energy Technologies. 2014;**2**(2):118-121

[97] Kavitha S, Adish Kumar S, Yogalakshmi KN, Kaliappan S, Rajesh Banu J. Effect of enzyme secreting bacterial pretreatment on enhancement of aerobic digestion potential of waste activated sludge interceded through EDTA. Bioresource Technology. 2013;**150**:210-219. DOI: 10.1016/j.biortech.2013.10.021

[98] Liu C, Zhang P, Zeng C, Zeng G, Xu G, Huang Y. Feasibility of bioleaching combined with Fenton oxidation to improve sewage sludge dewaterability. Journal of Environmental Sciences (China). 2015;**28**:37-42. DOI: 10.1016/j.jes.2014.05.039

[99] Fang W et al. Effect of alkaline addition on anaerobic sludge digestion

with combined pretreatment of alkaline and high pressure homogenization. Bioresource Technology. 2014;**168**:167-172. DOI: 10.1016/j.biortech.2014.03.050

[100] Li H, Jin Y, Mahar RB, Wang Z, Nie Y. Effects and Model of Alkaline Waste Activated Sludge Treatment. Bioresource Technology. 2008;**99**(11):5140-5144. DOI: 10.1016/j.biortech.2007.09.019

[101] Xu G, Chen S, Shi J, Wang S, Zhu G. Combination treatment of ultrasound and ozone for improving solubilization and anaerobic biodegradability of waste activated sludge. Journal of Hazardous Materials. 2010;**180**(1-3):340-346

[102] Chacana J et al. Effect of ozonation on anaerobic digestion sludge activity and viability. Chemosphere. 2017;**176**:405-411. DOI: 10.1016/j.chemosphere.2017.02.108

[103] Wang L. Different Pretreatments to Enhance Biogas Production: A comparison of thermal, chemical and ultrasonic methods. Sweden: Halmstad University. Master thesis P-51; 2011

[104] Wang LL et al. PH dependence of structure and surface properties of microbial EPS. Environmental Science & Technology. 2012;**46**(2):737-744. DOI: 10.1021/es203540w

[105] Mussoline W, Esposito G, Giordano A, Lens P. The anaerobic digestion of rice straw: A review. Critical Reviews in Environmental Science and Technology. 2013;**43**(9):895-915. DOI: 10.1080/10643389.2011.627018

[106] Modenbach AA, Nokes SE. The use of high-solids loadings in biomass pretreatment—A review. Biotechnology and Bioengineering. 2012;**109**(6):1430-1442. DOI: 10.1002/bit.24464

[107] Mouneimne AH, Carrère H, Bernet N, Delgenès JP. Effect of saponification on the anaerobic digestion of solid fatty residues. Bioresource Technology. 2003;**90**(1):89-94. DOI: 10.1016/S0960-8524(03)00091-9

[108] Kim J et al. Effects of various pretreatments for enhanced anaerobic digestion with waste activated sludge. Journal of Bioscience and Bioengineering. 2003;**95**(3):271-275. DOI: 10.1263/jbb.95.271

[109] Yuan H, Chen Y, Zhang H, Jiang S, Zhou Q, Gu G. Improved bioproduction of short-chain fatty acids (SCFAs) from excess sludge under alkaline conditions. Environmental Science & Technology. 2006;**40**(6):2025-2029. DOI: 10.1021/es052252b

[110] Cacho Rivero JA, Madhavan N, Suidan MT, Ginestet P, Audic J-M. Enhancement of anaerobic digestion of excess municipal sludge with thermal and/or oxidative treatment. Journal of Environmental Engineering. 2006;**132**(6):638-644

[111] Kohanski MA, Dwyer DJ, Hayete B, Lawrence CA, Collins JJ. A common mechanism of cellular death induced by bactericidal antibiotics. Cell. 2007;**130**(5):797-810. DOI: 10.1016/j.cell.2007.06.049

[112] Zhen G, Lu X, Kato H, Zhao Y, Li YY. Overview of pretreatment strategies for enhancing sewage sludge disintegration and subsequent anaerobic digestion: Current advances, full-scale application and future perspectives. Renewable and Sustainable Energy Reviews. 2017;**69**:559-577. DOI: 10.1016/j.rser.2016.11.187

[113] Zhang W, Yang P, Yang X, Chen Z, Wang D. Insights into the respective role of acidification and oxidation for enhancing anaerobic digested sludge dewatering performance with Fenton process. Bioresource Technology.

2015;**181**:247-253. DOI: 10.1016/j.biortech.2015.01.003

[114] Anjum M, Al-Makishah NH, Barakat MA. Wastewater sludge stabilization using pre-treatment methods. Process Safety and Environment Protection. 2016;**102**:615-632. DOI: 10.1016/j.psep.2016.05.022

[115] ErdenKaynak G, Filibelt A. Assessment of Fenton process as a minimization technique for biological sludge: Effects on anaerobic sludge bioprocessing. Journal of Residuals Science and Technology. 2008;**5**(3):151-160

[116] Kameswari KSB, Kalyanaraman C, Thanasekaran K. Effect of ozonation and ultrasonication pretreatment processes on co-digestion of tannery solid wastes. Clean Technologies and Environmental Policy. 2011;**13**(3):517-525. DOI: 10.1007/s10098-010-0334-0

[117] Carballa M, Manterola G, Larrea L, Ternes T, Omil F, Lema JM. Influence of ozone pre-treatment on sludge anaerobic digestion: Removal of pharmaceutical and personal care products. Chemosphere. 2007;**67**(7):1444-1452. DOI: 10.1016/j.chemosphere.2006.10.004

[118] Ariunbaatar J, Panico A, Frunzo L, Esposito G, Lens PNL, Pirozzi F. Enhanced anaerobic digestion of food waste by thermal and ozonation pretreatment methods. Journal of Environmental Management. 2014;**146**:142-149. DOI: 10.1016/j.jenvman.2014.07.042

[119] Szwergold BS. Maillard reactions in hyperthermophilic archaea: Implications for better understanding of non-enzymatic glycation in biology. Rejuvenation Research. 2013;**16**(4):259-272. DOI: 10.1089/rej.2012.1401

[120] Na SH, Shon HK, Kim JH. Minimization of excess sludge and cryptic growth of microorganisms by alkaline treatment of activated sludge. Korean Journal of Chemical Engineering. 2011;**28**(1):164-169. DOI: 10.1007/s11814-010-0334-0

[121] Rafique R, Poulsen TG, Nizami AS, Asam Z-u-Z, Murphy JD, Kiely G. Effect of thermal, chemical and thermo-chemical pre-treatments to enhance methane production. Energy. 2010;**35**(12):4556-4561. DOI: 10.1016/j.energy.2010.07.011

[122] Doğan I, Sanin FD. Alkaline solubilization and microwave irradiation as a combined sludge disintegration and minimization method. Water Research. 2009;**43**(8):2139-2148. DOI: 10.1016/j.watres.2009.02.023

[123] Weiland P. Biogas production: Current state and perspectives. Applied Microbiology and Biotechnology. 2010;**85**(4):849-860. DOI: 10.1007/s00253-009-2246-7

[124] Pérez-Elvira S, Fdz-Polanco M, Plaza FI, Garralón G, Fdz-Polanco F. Ultrasound pre-treatment for anaerobic digestion improvement. Water Science and Technology. 2009;**60**(6):1525-1532

[125] Bordeleau ÉL, Droste RL. Comprehensive review and compilation of pretreatments for mesophilic and thermophilic anaerobic digestion. Water Science and Technology. 2011;**63**(2):291-296. DOI: 10.2166/wst.2011.052

[126] Mottet A, Steyer JP, Déléris S, Vedrenne F, Chauzy J, Carrère H. Kinetics of thermophilic batch anaerobic digestion of thermal hydrolysed waste activated sludge. Biochemical Engineering Journal. 2009;**46**(2):169-175. DOI: 10.1016/j.bej.2009.05.003

[127] Yang X, Wang X, Wang L. Transferring of components and energy

output in industrial sewage sludge disposal by thermal pretreatment and two-phase anaerobic process. Bioresource Technology. 2010;**101**(8):2580-2584. DOI: 10.1016/j.biortech.2009.10.055

[128] Escamilla-Alvarado C, Ríos-Leal E, Ponce-Noyola MT, Poggi-Varaldo HM. Gas biofuels from solid substrate hydrogenogenic-methanogenic fermentation of the organic fraction of municipal solid waste. Process Biochemistry. 2012;**47**(11):1572-1587. DOI: 10.1016/j.procbio.2011.12.006

[129] Lu J, Gavala HN, Skiadas IV, Mladenovska Z, Ahring BK. Improving anaerobic sewage sludge digestion by implementation of a hyper-thermophilic prehydrolysis step. Journal of Environmental Management. 2008;**88**(4):881-889. DOI: 10.1016/j.jenvman.2007.04.020

[130] Ma J, Duong TH, Smits M, Verstraete W, Carballa M. Enhanced biomethanation of kitchen waste by different pre-treatments. Bioresource Technology. 2011;**102**(2):592-599

[131] Müller JA. Prospects and problems of sludge pre-treatment processes. Water Science and Technology. 2001;**44**(10):121-128

[132] Park C, Lee C, Kim S, Chen Y, Chase HA. Upgrading of anaerobic digestion by incorporating two different hydrolysis processes. Journal of Bioscience and Bioengineering. 2005;**100**(2):164-167. DOI: 10.1263/jbb.100.164

[133] Cesaro A, Belgiorno V. Pretreatment methods to improve anaerobic biodegradability of organic municipal solid waste fractions. Chemical Engineering Journal. 2014;**240**:24-37. DOI: 10.1016/j.cej.2013.11.055

Chapter 3

Composition, Production, and Treatment of Sewage Sludge

Rodrigo de Freitas Bueno

Abstract

Sewage treatment ultimately culminates in the concentration of the solid phase. Sludge are separated mainly in primary or secondary decanters. Even in biological treatment, where biological degradation of organic matter actually occurs, there is the separation of excess sludge concentrated in the bottom of the secondary settlers of activated sludge systems or biological filters. In fact, the production of sludge is an important differential in the choice of the treatment system. While purely aerobic systems such as activated sludge or high-rate biological filters can produce 0.6–0.8 kgSS/kgBOD applied, sludge production in an upflow anaerobic sludge blanket (UASB) reactor is only about 0.2 kgSS/kgCOD applied. Even the mixed anaerobic/aerobic system leads to less sludge production than that of an exclusively aerobic system. This advantage is very important nowadays, especially since, besides reducing the treatment needs, the difficulties with the final disposal of the sludge are usually very large.

Keywords: sewage sludge, densification, digestion, anaerobic sludge digestors, dimensioning

1. Introduction

At first, the main concern in relation to sewage treatment is to solve the problem of the liquid phase, leaving in the background the solution of the problem of the sludge generated in the treatment of this liquid phase. When there is a sewage treatment in which the generation of sludge is quite significant and the main concern with this sludge, at first, is restricted to its stabilization and dewatering to reach a solid content of the sludge in the range of 15–40%, aiming almost exclusively at its removal from the wastewater treatment plant (WWTP) area by trucks, however, has no clear definition of its final destination. In many cases, it is difficult to find suitable areas and conditions for the disposal of these sludge generated in the WWTPs [1]. Thus, the choice of the solution to be given to the sewage problem must also consider the solution to be given to the sludge generated in the treatment of the liquid phase. In order to study the alternatives for the treatment and disposal of sewage and the sludge generated in the treatment of the liquid phase, it is necessary to first know the possibilities of treating a sewage, in view of the quality of the final effluent to be obtained, as well as the quantities and qualities of the sludge produced in the WWTPs, which constitutes the main objectives of this work. In addition, an estimate of the costs of implementing WWTPs is presented for various types of sewage treatment systems. Currently, a very serious environmental problem, observed in metropolitan regions and medium-sized cities that have implemented sanitary sewage treatment systems, is related to the destination of the sludge produced in their WWTPs [2].

For an adequate solution of this problem, it is necessary initially to know the sludge production according to the used scrubber system. In terms of sludge production, considering only the quantitative aspect of solids produced in a WWTP, this work presents "per capita" values that can be useful to the technicians involved with the problem of the sludge destination considered here. However, the solid content (or moisture) of the sludge is another important parameter for the final disposal of the sludge and depends fundamentally on the type of stabilization used (anaerobic or aerobic biological or chemical) and the type of dewatering equipment used. In principle, the following ranges of solid content values should be considered [1, 3]:

- Anaerobically digested sludge, dehydrated by:
    - Plate press filter—solid content from 30 to 40%
    - Belt press filter—solid content from 16 to 25%
    - Centrifuges—solid content from 25 to 30%
    - Drying beds—solid content from 20 to 30%
- Aerobically digested sludge, dehydrated by:
    - Plate press filter—solid content from 25 to 35%
    - Belt press filter—solid content from 13 to 18%
    - Centrifuges—solid content from 20 to 25%
    - Drying beds—solid content from 20 to 30%

For the design of the new WWTPs, it is recommended that, in addition to the quality of the effluent to be required depending on the receiving body, the destination to be given to the sludge is also considered, as this aspect may be preponderant for the definition of the sewage treatment system and sludge to be adopted.

2. Sludge treatment steps

Sludge treatment can be subdivided into three main stages, although depending on the sewage treatment system adopted, some of them can be suppressed. This is the case, for example, with the activated sludge system with prolonged aeration, where the process operates in a range where the digestion of excess sludge can be dispensed with. Sludge discharged from upflow anaerobic sludge blanket (UASB) reactors also requires no densification and complementary digestion. Sludge densification may not be mandatory in activated sludge systems or aerobic biological filters, but, except in small systems, its inclusion is made possible by the benefits brought to later sludge treatment units [1]. When sewage treatment is carried out by pond processes, then the system operates in such a way that the sludge thickens and digests at the bottom of the stabilization or settling ponds, in the case of mechanically aerated pond systems. The problem becomes how to produce sludge removal mechanisms for final dehydration before being sent for disposal [2]. Depending on the use to be made of the sludge to be removed from the sewage treatment station, other treatment steps may be necessary, such as its disinfection for application on agricultural soil [4].

The objective of sludge densification is to reduce its moisture content, remove water and thus volume, and increase the solid content. The sludge discharged from "secondary" settlers of activated sludge systems with prolonged aeration has a solid content of less than 1%, and when a densifier raises it to 2%, there is a reduction in the sludge volume of 100% to be dehydrated. In a system of conventional activated sludge or aerobic biological filters, the sludge is mixed, primary and secondary. It is generated with a solid content between 1.0 and 1.5%, and its increase to about 4% allows an even greater reduction in volume, being able to prove the advantage of incorporating it into the system, in view of the much smaller required volume of anaerobic sludge digesters [1, 5, 6].

The purpose of sludge digestion is to complement its biochemical stabilization, that is, to increase the degree of mineralization. The sludge generated in conventional activated sludge systems and aerobic biological filters has a high volatile suspended solid/total suspended solid (VSS/TSS) ratio (e.g., 0.8) and thus does not allow good conditions for natural or mechanical dehydration. As what prevails in a sludge digestion stage is endogenous metabolism with destruction of VSS, this ratio is reduced (e.g., to 0.4), and the more mineralized sludge has better conditions for final dehydration. The objective of final dehydration is to remove water in order to achieve solid content above 20%, thereby drastically reducing the volume of sludge to be transported and making it compatible with applications such as disposal in landfills or agriculture [4, 6, 7].

2.1 Sludge densification

Sludge densification can be done by three main alternative processes. Gravity densification is applicable to both primary settler sludge and secondary settler sludge, that is, excess biological sludge, as well as to mixed primary and secondary sludge. Flotation densification with dissolved air can be an interesting alternative for the densification of excess biological sludge [3, 6, 8]. They result in solid contents higher than that of the gravity-densified sludge, and higher sludge loads per surface area of densifiers can be applied, resulting in the need for smaller areas of densifiers. The structure, however, is much more complex. Part of the final effluent from the WWTP, that is, from the treated sewage, feeds into the pressurization tank where the air is injected and, at a pressure of 4.0 kgf/cm^2, dissolves in the liquid in the form of microbubbles. Then, it is mixed with the sludge at the entrance from the bottom of the flotation chamber, with scraping removal of the densified sludge at the top and the sub-liquid to return to the entrance of the WWTP. Recently, machines for the mechanical thickening of sludge have been developed [5, 7]. They are machines designed to provide only a partial dehydration of the sludge, around 4–5%, for later final dehydration that can also be mechanized. Recent research has demonstrated the possibility of obtaining interesting advantages through the chemical conditioning of sludge prior to its densification. In this text, greater emphasis will be given to gravity densification [2].

2.2 Gravity density

Gravity densifiers are units similar to circular planter decanters, being fed with sludge through the center and at the top, inside a bulkhead that directs it to the bottom, from where it is removed after undergoing densification. Meanwhile, the supernatant liquid flows through the perimeter spillways positioned on the surface of the condenser and can be recirculated at the entrance of the WWTP [1, 3, 9]. **Figure 1** shows a photograph of an empty gravity densifier.

2.2.1 Dimensioning of gravity densifiers

The main dimensioning factor of gravity densifiers is the rate of application of solids, which is the mass flow of solids applied per unit of surface area of the densifiers. It depends on the type of sludge to be densified. The following ranges of values are proposed (**Table 1**).

The rate of application of solids should be the most restrictive factor in determining the surface area of the densifiers; however, the flow rate, sludge flow applied per unit of the surface area of the densifiers, must be kept within a certain limit, that is, $f_A = F/S_A < 16\ m^3/m^2$ day. Additional recommendations, such as minimum useful depth of densifiers equal to 3.0 m, maximum hydraulic holding time of 24 h, and mandatory mechanized sludge removal when diameters are greater than 3.0 m, are used. The maximum time limit for sludge retention in the compactor is characterized when there is a possibility of anaerobic decomposition and giving off of bad odors. When holding times greater than 24 h result from dimensioning, a portion of the treated sewage, the final effluent from the WWTP, can be recirculated, with a flow calculated to ensure compliance with the limit value.

Sizing example.

Data:

- Sludge type: primary + activated sludge.
- Sludge production: $\Delta X = 2254$ kg SS/day.
- Specific sludge mass: 1020 kg/m^3.
- Sludge solid content: 1%.
- Sludge flow: $F = 2254 \div (0.01 \times 1020) = 221\ m^3$/day.

Figure 1.
Sludge densifier by gravity. WWTP-Sabesp, São Paulo, Brazil (source: the author).

Sludge Type	Solid Application Rate (kgSS/m².day)	Solids content in the sludge Dense (%)
Primary	100 – 150	6 – 12
Biological Filter	40 – 50	4 – 10
Activated Sludges	20 – 40	1,5 – 4,0
Primer + Biological Filter	60 – 100	4 – 10
Primary + Activated Sludge	40 - 80	3 - 10

Table 1.
Application rate of solids according to the type of sludge. Source: adapted of Metcalf & Eddy, Inc. [3].

Solid application rate: 60 kgSS/m^2 day (adopted).
Required density area:

$$A_{\text{densifier}} = 2254 \div 60 = 37.6\ \text{m}^2\ (\text{diameter D} = 6.9\ \text{m}) \tag{1}$$

Adopting the diameter D = 7 m, the area of the condenser will be 38.5 m^2, and the resulting solid application rate will be 2254 ÷ 38.5 = 58.5 kgSS/m^2 day.
Adopted useful depth: Hu = 4 m.
Resulting useful volume: Vu = 4 × 38.5 = 154 m^3.
Hydraulic retention time: HRT = V ÷ F = 154 ÷ 221 = 0.7 day = 16.7 h.
Runoff rate: $f_A = F \div S_A$ = 221 ÷ 38.5 = 5.7 m^3/m^2 day.
Solid content in dense sludge: 4% (estimated).
Flow of thick sludge, for specific mass of 1030 kg/m^3:
$F_{\text{Sludge.Densifier}}$ = 2254 ÷ (0.04 × 1030) = 54.7 m^3/day.
Recirculation flow: 221–54.7 = 166.3 m^3/day.
The compactor must have mechanized sludge remover.

3. Sludge digestion

Depending on the type of sewage treatment and its operational conditions, the sludge from the solid separation units may require complementary biochemical stabilization. For that, aerobic or anaerobic digesters can be used; in both cases the reduction of sludge volatile content via endogenous metabolism is desired. Aerobic digesters are tanks equipped with an aeration system such as those used in activated sludge reactors. In view of the faster growth of these microorganisms in relation to anaerobes, it can be understood that the volumes of reactors are relatively smaller, which may lead to lower implantation costs than others, even though an aeration equipment is required [4]. However, for the aeration of concentrated sludge, the consumption of electric energy is quite high, and the difference between operating costs has led to the widespread use of anaerobic digesters in activated sludge systems and aerobic biological filters [7]. For this reason, a greater emphasis will be placed here on anaerobic sludge digestion.

3.1 Anaerobic sludge digestors

Anaerobic digestion can be considered as:

a. Conventional when the VSS application rate on the digester is equal to or less than 1.2 kg/m^3 day.

b. High rate when the rate of application of VSS on the digester is greater than 1.2 kg/m^3 day and equal to or less than 4.8 kg/m^3 day.

It is admitted to obtain a typical destruction of up to 50% of VSS and never exceeding 60%, according to the digestion and design conditions.

When selecting the VSS application rate, the influence of the internal temperature of the digester must be considered, and the need for heating of the unit must be verified. Anaerobic digestion should preferably be processed in the temperature range of 30–35°C (mesophilic digestion) or in the range of 50–57°C (thermophilic digestion) [1]. Lower temperatures result in less efficient digestion, to be considered in the project. High-speed primary digesters, with a VSS application rate equal to or greater than 0.5 kg/m^3 day, must be homogenized by one of the following devices: turbine mixer, gas homogenization system, and sludge recirculation pumps.

Digestion time should be:

a. For non-homogenized digester, ≥45 days.

b. For homogenized conventional digester, ≥30 days.

c. For high-rate digester, ≥18 days.

Second-stage digesters can be used for sludge storage and supernatant removal. The volume of the second-stage digester is about 1/3 of the volume of the first stage, determined by the criteria presented. **Figure 2** shows a photograph of the set of anaerobic digesters from WWTP.

Example of sizing a low-rate anaerobic digester.

Data:

- Sludge production: ΔX = 2254 kg SS/day.
- Volatile fraction: ΔXv = 1757 kg VSS/day.
- Sludge solid content: 4%.
- Sludge flow: F = 54.7 m^3/day.

VSS application rate: 0.5 kg VSS/m^3 day (adopted).

Figure 2.
Anaerobic digester. WWTP-Sabesp, São Paulo, Brazil (source: the author).

Required volume of anaerobic digesters:

$$V_{\text{Anaerobic.digestor}} = 1757 \div 0.5 = 3514\ m^3 \quad (2)$$

To meet the minimum detention time of 45 h, we have:

$$V_{\text{Anaerobic.digestor}} = 45 \times 54.7 = 2461.5\ m^3 \quad (3)$$

Three digesters of 1200 m^3 each should be adopted, making a total useful volume of 3600 m^3.

4. Sludge dehydration

The purpose of sludge dehydration is to raise the solid content generally above 20%, in order to reduce the volume to be transported and to allow its final disposal in landfills, agriculture, etc. It can be done naturally or mechanically. Natural drying can be done in drying beds or mud ponds. Sludge ponds should not be considered as a very suitable final solution. The sludge drying beds should be used more advantageously in small treatment systems. The required bed area is relatively large, on the order of 0.1 m^2 per inhabitant [2]. The cost of its structure, the operational difficulties with the removal of dehydrated sludge, and the excessive presence of rainwater can make its use unfeasible, especially in large treatment systems. Dewatering machines have grown in use in recent years, mainly plate press filters or conveyors, as well as centrifugal decanters. Despite the relatively high cost of these machines, the operational ease has enabled their adoption. For application of the machines, the previous conditioning of the sludge is required. In Brazil, chemical conditioning with ferric chloride and lime is common practice, a progressive practice of substitution of the use of polyelectrolytes [7, 8].

4.1 Sludge drying beds

The sludge drying beds are structures composed of bricks, arranged two by two and with joints filled with coarse sand. Under the bricks, layers of coarse sand and gravel of increasing granulometry are placed toward the bottom, an impermeable slab from which the liquid that infiltrates is drained and returned to the entrance of the WWTP. The sludge is disposed on the bricks, drying by infiltration of water in the bed and by evaporation in the sun. The beds are fed on a rotating basis, from channels with gates. A typical drying bed operation cycle, commonly adopted in projects in our region, is 30 days in total, with 20 days reserved for dewatering the sludge and 10 days for removing the dry sludge and rearranging the bed. Solids in the dehydrated sludge above 30% can be expected. For the determination of the necessary drying bed area, an application rate of solids as a criterion which cannot exceed 15 kg SS/m^2 × cycle is recommended. The rates in the region of 10–12 kg SS/m^2 × cycle are used in our region (Brazil). Once the total bed area is determined, it is subdivided into a number of beds that should not be too large to reduce operational difficulties [3]. **Figure 3** shows a view of the sludge drying beds.

Example of sizing sludge drying beds.

Data:

- Sludge production: $\Delta X = 2254$ kg SS/day.
- Volatile fraction: $\Delta Xv = 1757$ kg VSS/day.

Figure 3.
Drying beds. WWTP-Sabesp, São Paulo, Brazil (source: the author).

Considering that, prior to drying, the sludge will suffer a 55% reduction in volatile solids due to anaerobic digestion, we have:

VXv, RED = 0.55 × 1757 = 966 kg VSS/day.
ΔXp/drying = 2254–966 = 1288 kg SS/day.
ΔXp/drying = 1288 × 365 = 470,120 kg SS/year.
Adopting 12 drying cycles per year, we have:
ΔXp/drying = 470,120 ÷ 12 = 39,177 kg SS/cycle.

Adopting the rate of 12.5 kg SS/m^2 × cycle, we have the following required area of sludge drying beds:

$$A_{dry.sludge} = 39{,}177 \div 12.5 = 3134\ m^2 \quad (4)$$

Twenty-seven sludge drying beds of 6.0 × 20.0 m in dimensions in each plant are proposed. **Figure 4** shows an example schematic of a sludge drying bed.

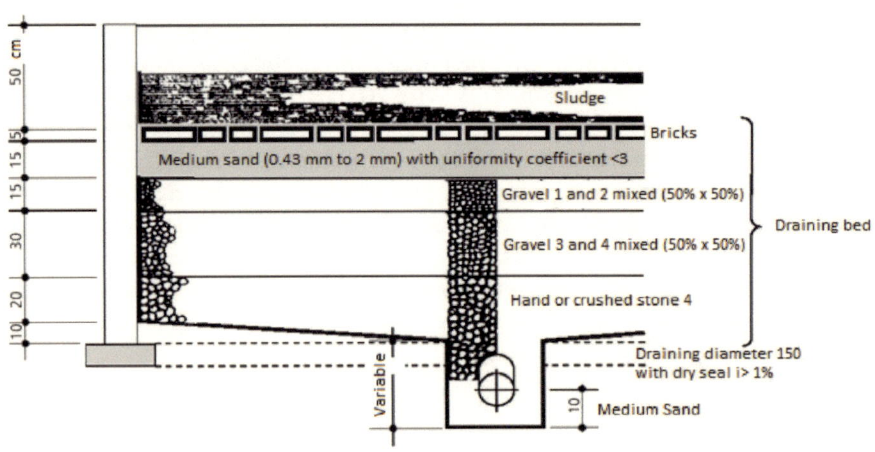

Figure 4.
Schematic cut of a sludge drying bed (source: the author).

4.2 Mechanical sludge dehydration

The main types of machines available on the market are plate press filters, continuous belt press filters, vacuum filters, and centrifugal decanters. When plate filter presses are used in activated sludge stations, dosages of ferric chloride, $FeCl_3$, of 7 kg/100 kg SS and hydrated lime, $Ca(OH)_2$, of 15 kg/100 kg SS are required, which makes this stage of the WWTP expensive, increases the volume of sludge, and hinders the agricultural disposal of sludge. When only polyelectrolytes are used, dosages of the order of 0.5–0.6 kg/100 kg SS are normally required [1, 2, 8]. The type of sludge conditioning and the dosages depend fundamentally on the state in which the sludge is generated, mainly its degree of mineralization, with less mineralized sludge being more difficult to dehydrate. **Figures 5** and **6** shows the diagram of the operation of a plate filter press.

Example of sizing a plate filter press

$$V = 100\ (SS) \div N \times P \times \rho \tag{5}$$

where V, volume of the filter press (L); (SS), suspended solids load (kg/day); N, number of presses per day; P, cake solid content (%); ρ, cake specific mass (kg/L).

Data:

$\Delta X = 6825$ kg SS/day.
$N = 4$.
$P = 30\%$.
$\rho = 1.06$.

Volume of the filter press:

$$V = (100 \times 6825) \div (4 \times 30 \times 1.06) = 5366\ \text{L} \tag{6}$$

Using plates (1.5 m × 1.5 m) of 3 cm thick, we have:

$$V_{\text{accumulated}} = 1.5 \times 1.5 \times 0.03 = 0.0675\ \text{m}^3 = 67.5\ \text{L} \tag{7}$$

$$\text{Number of plates} = 5366 \div 67.5 = 80 \tag{8}$$

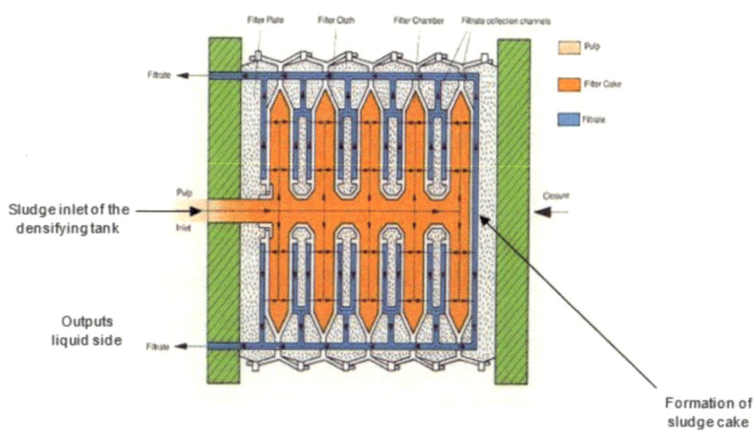

Figure 5.
Plate filter press (source: Andreoli [1]).

Figure 6.
Plate filter press in operation. WWTP-Sabesp, São Paulo, Brazil (source: the author).

Example of sizing a continuous belt filter press.
Data:

- ΔX = 8212 kg SS/day.
- ρ = 1030 kg/m^3.
- Solid content: 5%.
- Sludge flow: Q_{SLUDGE} = 8212 ÷ (0.05 x 1030) = 160 m^3/day.

Using the application rate of 300 kg SS/m × h and two filters with 1 m belt width, we have the following number of daily operating hours:

$$\text{No. of hours} \div \text{day} = 8212 \div (300 \div 2) = 14. \quad (9)$$

Polyelectrolyte consumption:

Medium: 6 kg/1000 kg SS.
Maximum: 8 kg/1000 kg SS.

Solid content in dehydrated sludge: 30%.
Volume of dehydrated sludge, with ρ = 1060 kg/m^3 and 90% solid capture.

$$\text{Dry sludge} = (0.9 \times 8212) \div (0.3 \times 1060) = 23\ m^3/\text{day}. \quad (10)$$

Figure 7 shows a continuous belt filter press.

4.3 Alternative design of a centrifugal decanter

Choosing a centrifuge with a feeding capacity of 10 m^3/h, we have the following number of daily operating hours:

$$\text{No. of hours} \div \text{day} = 160 \div 10 = 16 \quad (11)$$

Figure 7.
Belt filter press. WWTP-Sabesp, São Paulo, Brazil (source: the author).

Considering the dehydrated sludge at 20% solids, ρ = 1060 kg/m^3, and 90% solid capture, we have the following flow of dehydrated sludge:

$$\text{Dry sludge} = (0.9 \times 8212) \div (0.2 \times 1060) = 35\ \text{m}^3/\text{day} \qquad (12)$$

Figure 8 shows a centrifugal decanter.

4.4 Disinfection

Among the processes of sewage disinfection, chlorination and ultraviolet disinfection have been the most considered alternatives, with chlorination currently being the most economically interesting. Despite the potential for the formation of by-products, which can present toxicity, chlorination has been the solution used in almost all WWTPs, without using dechlorination for the time being. Another process used is heat treatment. This treatment effectively reduces pathogenic viruses,

Figure 8.
Photograph of a centrifugal decanter. WWTP-Sabesp, São Paulo, Brazil (source: the author).

bacteria, and helminth eggs to levels below those detectable. For the thermal inactivation of 99.9% of viable eggs in digested logos, an exposure time of 35 min at 58°C is required [3, 4].

5. Conclusion

This chapter covered the classic processes applied to the treatment, digestion, and drying of sewage sludge. Aerobic sewage treatment processes generate much more sludge than anaerobic systems. Of the sewage treatment systems, the stabilization ponds are the ones that generate the lowest amount of sludge, while conventional activated sludge systems have the highest volume of sludge to be treated. This is due to the fact that the sludge produced in the lagoons is retained for several years, undergoing digestion and densification, which induces a reduction in its volume.

Sludge digestion in the conventional activated sludge system is low due to the short sludge residence time in this system. The sludge filtration process leads to a higher concentration of solids than the thickening process. In filtrations with chemical conditions, the concentration of solids can increase in the order of 20–40% depending on the type of sludge and the form of filtration.

In the case of drying beds, it can be observed that in the period of 10–60 days of sludge rest, the concentration of solids increases to approximately 40%.

Aerobic digestion produces sludge with low dehydration capacity due to the destruction of the flake structure during the process of endogenous respiration that occurs in the aerobic digester. Anaerobic digestion can reduce the concentration of volatile solids in the sewage sludge by up to 60%.

The treatment, disposal, and reuse of WWTPs sludge are gaining more and more expression on the world stage, due to the increase in the number of installed sewage treatment plants and the need to meet the environmental requirements and currently the need for resource recovery natural.

Conflict of interest

The authors declare no conflict of interest.

Appendices and nomenclature

ΔX	sludge production
BOD	biochemical oxygen demand
COD	chemical oxygen demand
F	flow
S_A	surface area
SABESP	sanitation company of the state of São Paulo, Brazil
TSS or SS	total suspended solids
UASB	upflow anaerobic sludge blanket
VSS	volatile suspended solids
WWTP	wastewater treatment plant
Xv	volatile fraction

DOI: http://dx.doi.org/10.5772/intechopen.91665

Author details

Rodrigo de Freitas Bueno
Central of Engineering, Modelling and Applied Social Sciences of the Federal University of ABC, Santo André, SP, Brazil

*Address all correspondence to: rodrigo.bueno@ufabc.edu.br

References

[1] Andreoli CV, Von Sperling M, Fernandes F, Ronteltap M. Sludge Treatment and Disposal. IWA Publishing; 2007

[2] Bishop P. Municipal Sewage Sludge: Management, Processing and Disposal. CRC Press; 1995

[3] Metcalf & Eddy, Inc. Wastewater Engineering: Treatment, Disposal, Reuse. 3rd ed. New York: McGraw-Hill International Editions; 1999

[4] de Chernicharo CAL. Post-treatment options for the anaerobic treatment of domestic wastewater. Reviews in Environmental Science and Bio/ Technology. 2006;**5**(1):73-92

[5] Hudson JA, Lowe P. Current technologies for sludge treatment and disposal. Water and Environment Journal. 1996;**10**(6):436-441

[6] Van Haandel AC, Lettinga G. Anaerobic Sewage Treatment. London, England: John Wiley & Sons; 1994

[7] Nowak O. Optimizing the use of sludge treatment facilities at municipal WWTPs. Journal of Environmental Science and Health Part A. 2006;**41**(9): 1807-1817

[8] Dohányos M, Zábranská J, Jenícek P. Enhancement of sludge anaerobic digestion by using of a special thickening centrifuge. Water Science and Technology. 1997;**36**(11):145-153

[9] Shammas NK, Wang LK. Gravity thickening. In: Biosolids Treatment Processes. Humana Press; 2007. pp. 45-69

Chapter 4

Solid Waste Management: Current Scenario and Challenges in Bengaluru

B.P. Naveen and P.V. Sivapullaiah

Abstract

Municipal solid waste management (MSWM) has become one of the significant environmental issues, particularly in developing countries. Bengaluru, the state capital of Karnataka, is one of the fastest growing cities in Asia. The Bruhat Bengaluru Mahanagara Palike (BBMP) with an area of 2190 km^2 and a population of about 10.18 million generates around 5000 metric tons per day of solid waste at an average generation rate of 0.5 kg per capita per day (kg/capita/d). Presently, Bengaluru City is facing significant problems due to existing disposal practices of generated waste, incurring high cost due to lack of proper infrastructural facilities; also, the open dumping in the expanding zone of the city poses severe problems to the structures constructed on these old dumps. In the meantime, groundwater quality deteriorated due to improper leachate management. Intending to assess the possible impacts on the water environment and suggest a better waste management strategy, the present paper discusses the potential for handling the wastes, thereby reducing the amount of waste to be transported to the landfill. If this waste is used for energy and nutrient recovery, decentralization could also become commercially viable and address the technology-wise deficiencies in the existing MSWM system of Bengaluru City.

Keywords: collection, disposal, transportation, processing, management, strategy

1. Introduction

In ancient times, a harmonious and balanced relationship between humans and nature on this earth is necessary for livelihood. As civilization advanced, humans directly or indirectly interfered with the natural environment. This led to an imbalance in the human-nature relationship, finally leading to environmental problems like soil, air, and water pollution and accumulation of municipal solid waste (MSW).

In olden days, MSW disposal did not pose significant problems because the population was very less and the availability of land for the dumping of wastes was large. But these days MSWM is a serious problem everywhere. Due to rapid industrialization and increased population levels, the generation rate of MSW in metropolitan cities accelerates. This has led to the migration of people from villages to cities, which generates thousands of tons of MSW daily with rapid change in the quantity and character of the waste in line with the changing lifestyle of the people and also with the changes in the market technology, building technology, and fuel technology.

IntechOpen

The environmental degradation and energy crisis are two significant issues for global sustainable development. Due to rapid urbanization, industrialization and increase of the growth of population have led to severe substantial waste management problems in several developing countries like India, Malaysia, Nepal, and Bangladesh. As the village develops into towns and cities, in developing countries the disposal of solid waste onto access ways, empty lands, and waterways has been witnessed. Presently, more materials are consumed than required to meet their daily needs by a greedy human. Human beings generate domestic, agricultural, industrial, and medical wastes at every level of development. This waste comprises of both solid and semisolid organic wastes, which may be biodegradable and non-biodegradable. Hence, proper collection and subsequent disposal of waste assumed vital importance in community environmental sanitation programs.

MSW has become one of the significant environmental issues, particularly in developing countries. The solid waste generation mainly consists of biodegradable and non-biodegradable waste materials produced due to several societal activities. The improper dumping of solid waste pollutes the air, soil, and water. The BBMP with an area of 2190 km^2 and a population of about 10.18 million generates around 5000 metric tons per day of waste at an average generation rate of 0.5 kg per capita per day (kg/capita/d). Presently, Bengaluru is facing significant problems due to existing disposal practices of generated waste incurring high cost due to lack of proper infrastructural facilities; also the open dumping in the expanding zone of the city poses severe problems to the structures constructed on these old dumps. In the meantime, groundwater quality deteriorated due to improper leachate management [1].

In Bengaluru, there are more than 60 illegal dumpsites identified. While BBMP and the Karnataka State Pollution Control Board (KSPCB) close these dumpsites, the new ones emerge elsewhere, posing health risks to residents in their vicinity. The MSW (Management and Handling) Rules, 2000, recommend source-specific waste collection and transportation in addition to appropriate processing and disposal. There is a lack of knowledge of the quantity and characteristics of reliable waste aids in the preparation of a long-term plan for an MSWM system. So, it was deemed necessary by the BBMP to assess the current status of the municipal solid waste management system in Bengaluru [2].

In this context, the present study discusses the potential improvement in handling the wastes and reduces the amount of waste to be transported and dumped in the landfill. If this waste is used for energy and nutrient recovery, decentralization could also become commercially viable. Moreover, it also addresses the wise technology deficiencies in the existing MSWM system of Bengaluru.

2. Bengaluru scenario

The city of Bengaluru (12.98°N and 77.58°E) in Karnataka is the state capital, and it has a mild and salubrious climate. It is located at an elevation of 900 m. Since the 1980s, Bengaluru has enjoyed the reputation of being one of the fastest developing cities in Asia [3]. The Bengaluru Metropolitan Area covers an area of 1258 sq. km and is the fifth largest city in India. However, with an increased population level, rapid economic growth, and a rise in community living standard, the generation rate of MSW in metropolitan cities accelerates. The local authorities are struggling to provide the proper solid waste management system to a satisfactory level. Recently the authorities have taken initiatives and measures to organize the MSWM sector. This research would help to identify techniques suitable for the current scenario, the loopholes in the adopted methods, and the possible alternatives.

3. Municipal organization

The BBMP has a city council that consists of 123 elected members or councilors, each representing a ward. Both the mayor and deputy mayor are chosen from among councilors for a 1-year term. The BBMP has 15,000 employees and is headed by the commissioner. The commissioner is the head of the BBMP, appointed and deputed by the State Government of Karnataka and responsible for performing duties and functions.

Presently, the Bruhat Bengaluru Mahanagara Palike is the agency vested with responsibility for effective solid waste management system for the Bengaluru City. For a more efficient and effective approach, the Bengaluru City has been divided into different administrative units. There are 294 health wards within the BBMP. Presently, in Bengaluru, there are 198 such administrative or political wards (**Figure 1**). Within the BBMP, there are two departments, namely, the health department and engineering department. The health department is mainly responsible for the collection, transportation, and disposal of solid waste. The engineering department handles the removal of construction and demolition waste, while they also provide technical and infrastructural support to the health department.

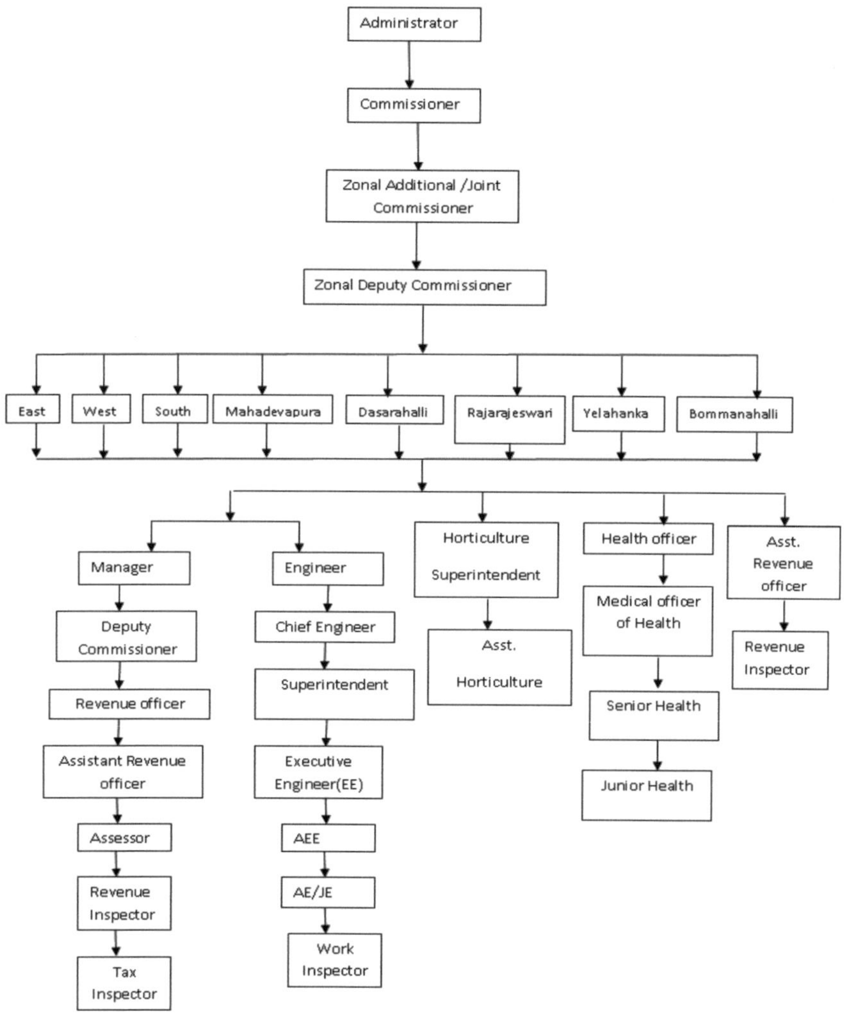

Figure 1.
BBMP Zonal organization chart.

4. Waste generation

At present, 10% of solid waste is recycled in Bengaluru. Most of the literature reported that the waste generation rate is 0.4–0.6 kg/capita/day. The proposed waste generation rate is about 0.5 kg/capita/day in Bengaluru [4]. Since 1990, the composition of India's urban wastes has changed drastically. At present, the waste generation is about 5000 metric tons, and waste generation is likely to grow over the coming years. Going by the present trend of increase in the quantity of waste, the waste quantity projected for the next 20 years is shown in **Figure 2**.

4.1 Waste composition

The changes in the composition of MSW should form essential criteria for any waste management system. Hence, the data available on the composition of the waste from different sources over the years have been collected and analyzed. **Figure 3** shows variation in MSW composition from 1999 to 2013 in Bengaluru City. With the increase in urbanization and change in food habits and lifestyle, the amount of MSW has been multiplying, and there is variation in waste composition.

The changes in the composition of MSW should form essential criteria for any waste management system. Hence, the data available on the composition of the waste from different sources over the years has been collected and analyzed. The data presented in **Table 1** was statistically analyzed to get the variations of a different type of waste in rapidly growing cities like Bangalore.

The MSW composition generated in Bengaluru has changed considerably from 1999 to 2013, which is evident in **Table 1**. It was observed that there was an increase in the biodegradable percentage in Bengaluru City from 42% in 1999 to 61% in 2013, indicating increased organic waste generation in the city, which may be primarily due to increasing population, improper solid waste management, or accumulation of green waste. Similarly, there was a 16% decrease in the paper, cardboard, and leather wastes in Bengaluru City, indicating recycling activities of paper and

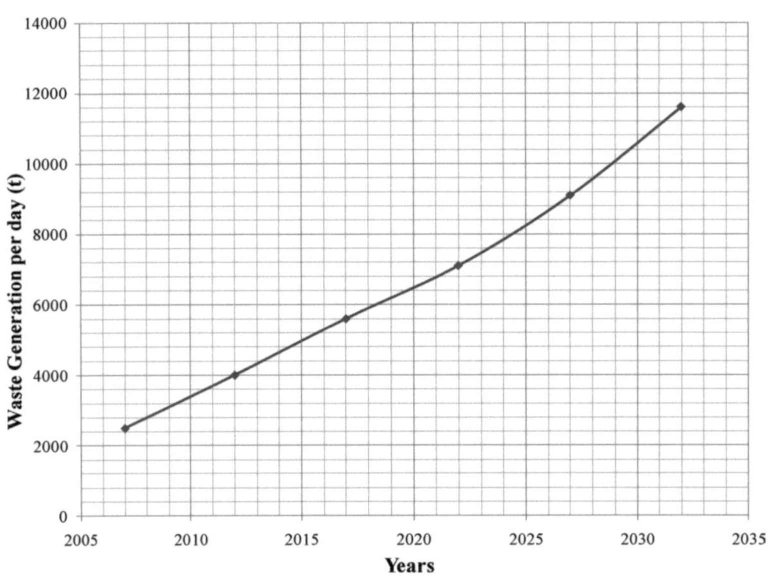

Figure 2.
Waste quantity expected for the next 20 years.

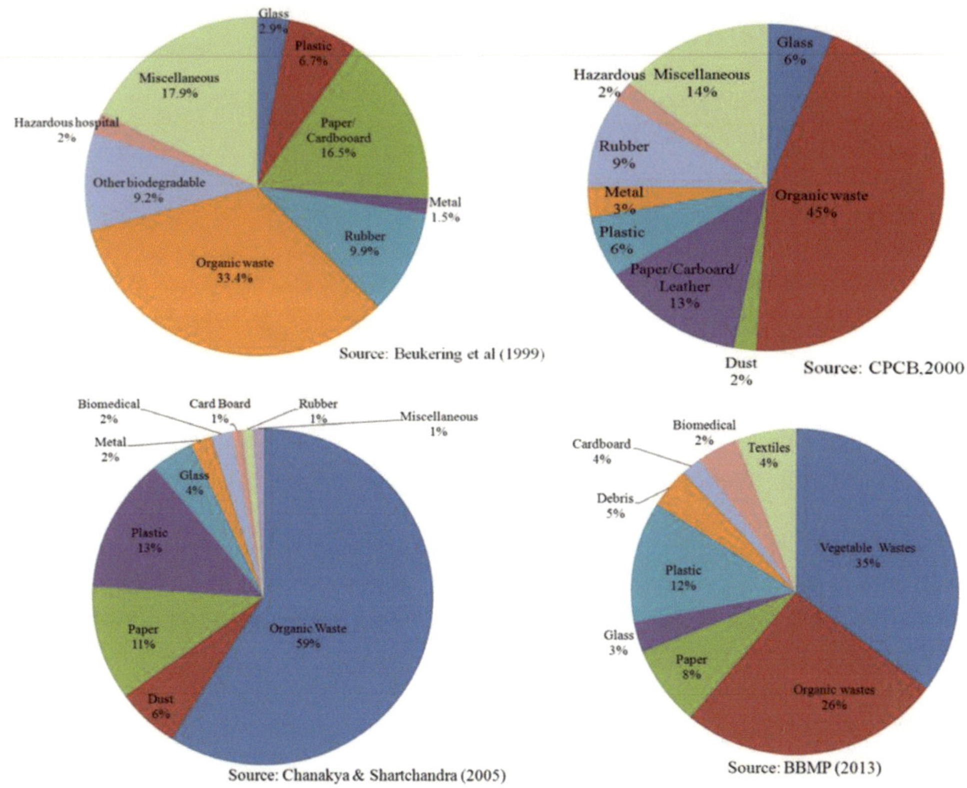

Figure 3.
Bangalore's urban waste composition.

Year	Biodegradable	Paper/leather/ cardboard	Rubber/ debris	Glass	Plastic	Metals	Others
1999	42	16	10	3	7	2	20
2000	45	13	9	6	6	3	18
2007	59	12	5	3	12	1	8
2013	61	9	4	1	7	1	16

Source: data extracted from Bangalore's Urban Waste Composition [4–7]
#All components of MSW are expressed in %.

Table 1.
Variations in MSW composition in Bangalore City from 1999 to 2013.

cardboard. A considerable increase in plastic wastes was also observed in 2007, which might be due to the urbanization and increased use of plastic carry bags. In 2013, the percentage of plastic waste decreased to 7%, which may be attributable to the effective ban of plastics carry bags below 40microns within city limits in 2012. The glass, metals, and rubber fraction observed a decrease in MSW composition, indicating the decreased use of glass and metal products and effective recycling of glass and metal products by segregation at sources itself.

The variations in MSW composition shown in **Figure 4** can be utilized in choosing the best method of MSW disposal in Bengaluru City. Biodegradable percentage (61% in 2013), more than 50%, suggests employment of methods, such as windrow composting, community composting process, pyrolysis, and vermicomposting, which assists in manure generation for agricultural practices; also, biomethanation

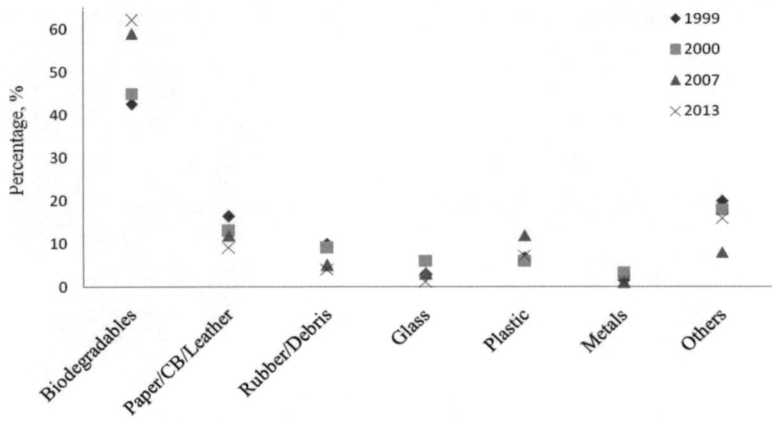

Figure 4.
Variations in MSW composition from 1999 to 2013 in Bangalore City.

can be employed to produce biogas, which can be utilized as a fuel or alternative source for electricity generation. As the significant composition of MSW is organic, waste treatment options like composting were successful in Bangalore, and 3.14% waste reduction was achieved through composting [8].

5. Collection and transportation of waste

The objective of solid waste management is to remove discarded materials from inhabited places promptly to prevent the spread of disease, to reduce esthetic results arising from purifying organic matter, and, equally important, to dispose of the discarded materials in a manner that is environmentally acceptable [9].

5.1 Collection

Currently, in Bengaluru, source segregation is still a concern through awareness in picking up slowly. BBMP handles about 30% of solid waste, and the remaining waste activity is outsourced (starting from primary collection to disposal). Solid waste collection is carried out in two phases. The first phase is a primary collection, in which the solid waste is collected on auto tipper and pushcarts. An auto tipper has been provided for every 1000 households and a cart for every 200 homes. About 20,000 pourakarmikas are being utilized (both BBMP and contractors) in the door to door collection, street sweeping, and transportation of MSW. The collected solid waste from houses is brought to a common point, i.e., secondary locations from where the waste is transferred to landfill sites/treatment through tipper lorries and compactors. **Figure 5** indicates a typical scheme of how the collection and transportation are being practiced in most of the wards.

This activity is assigned to self-help groups (SHGs), which are basically below poverty women's groups. BBMP has allocated 3197 pourakarmikas (sweepers) and 18,562 pourakarmikas from a contractor who performs door to door collection and sweeping activities. Annually about 250 crores are spent on solid waste management, i.e., BBMP pourakarmika salary, contract payment, and tipping fees (**Figure 6**).

The survey carried out in Malleswaram by [10] indicated that no norms/guidelines had been followed in setting up waste segregation practices adopted in this ward. There is a lack of awareness among people that leads to confusion. The six

Figure 5.
MSW collection system.

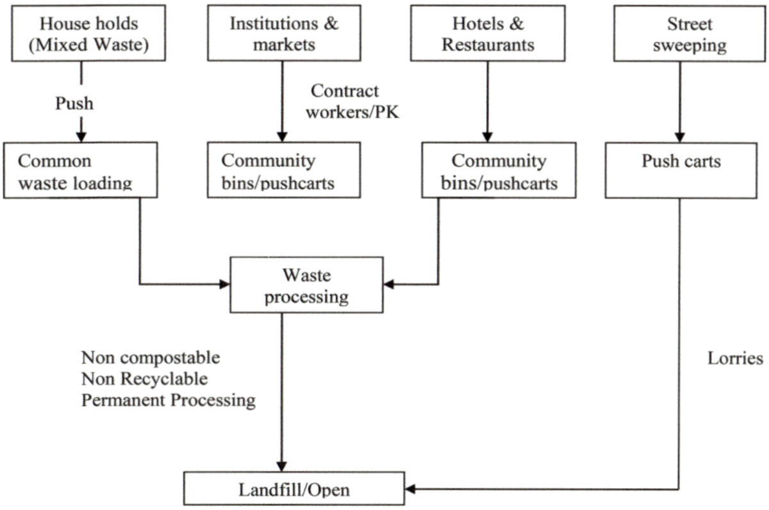

Figure 6.
MSW collection process in a typical residential/commercial area.

categories for segregation mandated by BBMP are overwhelming and are a deterrent to segregation. The waste collectors (pourakarmikas) lack training in proper segregation practices and its importance. Incomplete segregation is the predominant practice currently, and steps to realize a higher level of compliance and efficiency need to be effected.

5.2 Transportation

Transportation of waste from collection centers to the landfill site is another crucial step in waste management. At present, waste transportation is using pushcarts, auto, etc., which bring waste to primary collection centers. From there, trucks collect the municipal solid waste and transport it to waste disposal sites/landfill.

The issues in transporting waste are mentioned below:

- The waste spills from the trucks due to open beds in trucks and tractors, during transport, thereby causing a nuisance.
- Solid waste loaded manually in a truck without using the protective gears is dangerous to the health of workers.
- A secondary storage system is not well synchronized with the transport system. Problems arise when a transport fleet is modernized because waste at a secondary storage system is still dumped on the ground.
- The area cannot be appropriately serviced due to an inadequate number of vehicles.
- Due to inadequate workshop facilities and maintenance procedures, the vehicles are poorly maintained. This problem leads to a breakdown of trucks, and they become out of service for a long time.

5.3 Effective solid waste management

Based on the above-presented data and analysis, for effective solid waste management, the following suggestion can be made:

1. Establish the segregation at the source itself, and encourage by giving incentive to the contractors with the performance in segregation.
2. Establish the wet waste processing units for composting, and encourage decentralized processing for dry waste collection center at the source.
3. Set up the segregation units and zone-wise processing facilities to ensure 100% processing of municipal solid waste, and minimize the solid waste quantity going to the landfill sites.
4. Reduce transportation of municipal solid waste using the above measures of decentralized as well as zone-wise processing units.

6. Waste disposal practices in Bengaluru

Currently Bengaluru does not have any appropriate scientific treatment techniques for waste generated by municipal and industries around Bengaluru. This has led to the development of various unauthorized dumpsites. The solid waste, generated from hotels, restaurants, Kalyana mandapas, markets, etc., is being directly collected and transported to the treatment/disposal facilities. The treatment facilities have been developed around the city, and their spread over the zone is set out in **Table 2**.

7. Decentralized waste processing plant in Bengaluru

For making effective solid waste management services, it is essential to select appropriate technology, which can suit and work in the given area successfully.

Sl. No	Zone	Existing disposal site/facility
1	South	Bingipura, Mavallipura, KCDC
2	East	MSGP, Mavallipura, KCDC
3	West	Terra firma, Mavallipura, KCDC
4	Yelahanka	Mavallipura, Terrafirma
5	Bommanahalli	Bingipura, Laxmipura
6	Mahadevapura	Terrafirma
7	Dasarahalli	MSGP
8	Raja Rajeshwari Nagar	MSGP/Terrafirma

Table 2.
Existing treatment and disposal facilities in BBMP.

Simultaneously, proper measures have to be considered for institutional strengthening and internal capacity building. Institutional strengthening can be done by adequately decentralizing the administration, delegating adequate powers at the decentralized level, providing training to the existing staff, and assigning the responsibility for the workforce as well as for supervisory staff. NGO/private sector participation is necessary for making service competitive and efficient. The land is scarce, and public health and environmental resources are precious.

In this direction, many decentralized facilities are being established. The decentralization of administration has to be implemented in large cities to make solid waste management service effectively. Decentralization can be divided into three tiers—one at the ward level, second at the zone level, and third at the city level. The BBMP established the decentralized processing units for dry waste; details are as follows.

7.1 Dry waste collection center

Dry waste materials like batteries, tin cans, plastic-coated milk cartons, nylon, cigarette butts, and leather all take varying lengths of time to degrade: not less than 10 years. Hence, the best way to dispose of these dry waste items is to reuse/recycle them. Nearly 70% of all the dry waste thrown away can be safely disposed of this way.

The aim of Dry Waste Collection Centers (DWCCs) is to keep as much waste out of the landfills/waste dumps as possible and to help make waste useful and profitable. DWCCs run by various agencies in Bengaluru in coordination with the BBMP is decentralized bulk sorting and processing facilities. About 185 DWCCs have been established and functional. DWCCs are set up on municipal/government/private lands and various NGOs, waste pickers, and contractors; self-help groups have been involved for effective functioning. The dry waste generated in the wards is collected and further segregated and sent for recycling from these recycling centers. Receiving in bulk provides these informal sector workers with more significant returns and creates more jobs.

7.2 Sanitary landfill site

Currently, Bengaluru can handle the waste of about 2100 TPD. The existing capacity of waste treatment facility at Mavallipura is 600TPD, Karnataka Composting Development Corporation Ltd. (KCDC) is 300TPD, and Terra Firma is about 1500 TPD as shown in **Figure 7**. For achieving 100% processing of solid

Figure 7.
Sanitary landfill sites in Bangalore.

waste, the government has identified new landfill sites. These sites are being set up at the following locations: Kannahalli (500TPD), Seegihalli (200TPD), Doddabidarakallu (200TPD), Lingaderenahalli (200TPD), Subrayanpalya (200TPD), Chikkanagamangala (500TPD), and KCDC (upgradation) (500TPD). Majorly we waste composting plants with a provision to screen compost out of mixed MSW and provision also to store the non-compostable/non-recyclable materials. These materials can be used for co-incineration in cement industries/ power generation [4].

8. Emerging technologies for wet waste disposal

Once the solid waste is collected from the different sectors of the community, the next problem is regarding the safe, economical, and efficient disposal options. Suitable decisions have to be made in this regard to avoid illegal dumping and open dumping of solid wastes that are dangerous and a threat to the environment. Open burning of solid wastes releases smoke containing pollutants harmful to human health and the environment. Therefore, the community has to face severe inconveniences due to illegal dumping practices. Hence, conventional methods employed to safe disposal options of MSW include composting, waste to energy (such as biogas production and incineration), and landfilling. Incineration and composting of MSW are a standard solid waste treatment or processing methods, as they produce secondary waste such as non-biodegradable material rejects from composting and ash from incineration that needs to be disposed of further [11].

8.1 Aerobic composting

Aerobic composting process involves piling up of waste and requires regular turning, manually or by mechanical devices, and sufficient air, and oxygen has to be provided during the decomposition by bacteria, fungi, and microorganisms like actinomycetes. A mesophilic bacterium is an initial process, which oxidizes the organic matter to carbon dioxide and generates the heat and temperature rise to about 45°C. In the next process, thermophilic bacteria continue the decomposition; in this phase, temperature further rises to about 60°C. Three weeks is required for stabilized compost and fall in temperature of the compost mass. The final product of the compost should have a dark brown color and earthy smell.

8.2 Windrow composting

The waste is dumped in the windrow platform; large items like woods, plastics, clothes, thermocol, etc. are removed; and inoculum will be sprayed on the waste. The inoculum will be prepared using the mixture of bacteria, cow dung, and water. The treated waste is then heaped in windrows with long rows approximately 2 meters in height and 3 meters in width; length will be depending on the size of the landfill site. There are seven rows, each row for each day of the week. Every week these rows are turned for 5 weeks. These rows are turned to remove moisture, improve porosity and oxygen content, and redistribute hotter and colder portions of the pile. As time passes, the sizes of the rows get reduced due to the decomposition of the waste and the resultant volume reduction. Hence, the number of final rows will be decreased than the number of initial rows. Composting will be completed in 25–30 days. This interval is known as maturation in which waste will undergo mechanical process operation. In mechanical processing, sieving occurs in three stages: in the first stage, sieve employs 36 mm mesh, the second stage applies 16 mm mesh, and the third stage has a 4 mm mesh, as shown in **Figure 8**. At each stage of sieving the reject, materials are separated and either reused or disposed of at the landfill. Any leachate or runoff created must be collected and treated. To avoid problems with leachate or runoff, waste piles can be placed under a roof, but doing so adds to the initial costs of the operation [11].

The following are the challenges in windrow composting:

1. Minor mechanical fault leads to a breakdown due to the unavailability of spare parts.

2. The major difficulty is due to the nature of waste. Pulverizers get frequently clogged with pieces of plastic, rubber, leathers, etc., and due to metal and glass pieces, the blades breakdown. If waste is mixed with soil, it causes a problem in the process, lowering the quality produced.

3. Lack of continuous power supply.

4. In the rainy season, the process cannot be carried out.

8.3 Community composting

In this process, daily wet waste is collected by the housekeeping staff, directly dumping into the tank. After filling, the tank is covered with a layer of refuse 15–20 cm deep. The materials are allowed to remain in the pit without turning and watering for 3 months. To keep the decomposers working, the aeration aid is needed during the initial pile construction. As long as plenty of air is available, aerobic decomposers work faster and more efficiently, providing you with finished compost on a more rapid time. Charcoal is placed in the tank. Hence, foul smell is avoided. It takes about 3 months to obtain the finished product. Community composting process is shown in **Figure 9**.

Figure 8.
Windrow composting process.

Figure 9.
Community composting process.

8.4 Biomechanical composting

In this process, organic waste such as vegetable and fruit peels and food leftovers, bones, meat, eggshells, household sweeping dry leaves, garden waste, cattle dung, etc. collected from the apartments and other places were identified for segregation for removal of plastic, glass, clothes, paper, leather, etc. for recycling purposes. After segregation of organic waste, it is then fed into the mechanical unit (i.e., organic waste converter) which converts this into a homogenized, crushed, odor-free output (**Figure 10**). The output goes to the curing system for stabilization. Aerobic microbial decomposition controls the entire process; the transition takes place from low pH levels to high pH levels and then stabilizes. This manure is free from weed, foul smell, and pathogen as the process is aerobic. This is environment-friendly operation; this system takes only 15 min to convert the organic waste into a homogenized output.

8.5 Vermicomposting

More than 50,000 populations of worms can support the moist compost heap of 2.4 m by 1.2 m and 0.6 m high. Organic residues such as straw and other crop residues, animal manure, green weeds, and leaves are filled in the pit and covered loosely with soil and kept moist for a weak. On the top of the heap, well-watered, the worms will be introduced, and air is provided for quick decomposition. *Lumbricus rubellus* (red worm) and *Eisenia foetida* are thermo-tolerant and are particularly useful for vermicomposting. Ideally, the compost pits were left for 2 months, and such pits should be shaded from hot sunshine and kept moistly. 1 kg of worms can produce 10 kg of castings within 2 months. Then the pit will be excavated to the extent of about two-thirds to three-quarters, and worms will be removed by hand (**Figure 11**). The remaining worms will be left in the pit itself for further composting with fresh organic residues. To get a good quality of compost

Figure 10.
Biomechanical process.

Figure 11.
Vermicomposting process.

material, sun-drying and sieving have to be carried out. The end product of compost is an ideal constitution and structure. For vermicompost, the unit has to be protected against chicken, other birds, rodents, and heavy rains.

The following are challenges in vermicomposting:

1. This concept is suitable for only small-scale applications and not an appropriate solution for large-scale application, e.g., 100–300MT/d capacity plants.

2. The exotic species are found to be costing between Rs. 500 and 1000/kg, and indigenous species of earthworms are not found useful.

3. The raw waste cannot be fed directly to earthworms, thus necessitating the pre-processing of waste to avoid toxicity.

4. Earthworms are so sensitive to temperature (ideally between 20 and 28°C); worms die due to heat built up in the rotting pile or summer.

9. Waste-to-energy technology

Some demonstrated technology approaches are available for waste-to-energy (WTE) projects today which are anaerobic processing/biogas production, refuse-derived fuel (RDF), and plasma gasification.

9.1 Biogas production/anaerobic processing

Biomethaization plants are being established for wet solid waste at 16 locations, out of which 8 have been made functional. 400 units are generated per day per plant. The biogas produced from the bio-mechanization of plants is being utilized to light the street lights in that locality.

Biogas is produced in the absence of oxygen or an anaerobic environment, due to the decomposition of organic material through certain bacteria. The whole process is referred as anaerobic digestion because biological decomposition takes place in a reactor, where bacteria produce biogas. This biomass can stay in the reactor for about 2–3 weeks. In the end, the by-product produced in this process is a solid residue that is high-grade manure. Generally, in the biogas plant, biomass like vegetable wastes and animal excreta undergo decomposition in the absence of oxygen and form a mixture of gasses. Biogas consists of about 2/3 methane (CH_4), 1/3 carbon dioxide (CO_2), a little hydrogen sulfide (H_2S), and a low hydrogen (H_2). It is created by the decomposition of manure and other forms of organic waste from households or industries in anaerobic tanks where it is heated (**Figure 12**). The biogas is used for cooking and lighting purposes.

The following are the disadvantages of anaerobic processing:

1. In a large industrial scale, this process is not very economical compared to biofuel.

2. It is challenging to increase the efficiency of biogas systems.

3. The gasses come out from biogas as impurities, which are corrosive to the metal parts of internal combustion engines.

4. Not feasible to locate at all the locations.

9.2 Refuse-derived fuel

RDF plants are in the initial stage of development in India. In this process, plenty of combustible components of MSW, such as plastic, cardboard, paper, and biodegradable waste, are converted into fuel pellets. It mainly involves drying, separation of combustion from MSW, size reduction, and pelletization after mixing

Figure 12.
Biomethanation process.

with binder and additives as required. If MSW contains 35–40% moisture content, then it involves air-drying for 2 days. Then the waste is spread, and manual inspection is carried out to remove large size debris, tires, tree stones, tree trunk, etc. The air-dried MSW is fed uniformly into a rotary drying system, i.e., hot air generation burning oversized wastes. 10–12% moisture content is suitable to be maintained in MSW for densifying into fuel pellets.

After air-drying, MSW is passed through screening equipment (below 8 mm) to separate heavier combustibles and ferrous materials; it may cause harm to process equipment. Fine fraction contains organic matter, and it is already proven to be useful as garden manure.

Air-dried waste is then passed through the density separator; here light combustibles and an air barrier separate dense fractions (e.g., stones, glass, etc.). Parallel MSW is passed over a magnetic separation unit to remove magnetic materials. The binder and additives are mixed with ground solid waste in the mixer before pelletizing. Once pellets are coming out from the pelletizer, they are cooled and stored for dispatch. The RDF pellets are used as a coal substitute at a lower price.

9.3 Plasma gasification

In plasma gasification, the process converts all types of wastes into a synthesis gas composed of hydrogen, nitrogen, carbon monoxide, and water. This synthesis gas can be used to generate electrical power and useful liquid fuel, such as ethanol.

The following are the advantages of plasma gasification:

1. No segregation of MSW needed.

2. Waste to energy.

3. A 120 megawatt (MW) facility will require, on average, consuming 3000 tons of garbage per day (Mt garbage/d).

4. The plasma system can be retrofit on the existing power-generating plants, reducing time, and greenhouse gas emission.

5. Plasma systems can use old landfills, thoroughly cleaning up and beautifying our landscape.

6. Plasma gasification is an affordable, cost-effective solution compared to other alternative energy solutions.

The following are the disadvantages of the plasma gasification:

1. Lack of regulations needed for permits

2. Financial risk

3. Technical risk

4. Economics

10. Options available for waste disposal

In this section are some of the options available for waste disposal. However, it should be noted that the option selected for waste disposal must mesh with the existing sociocultural milieu, infrastructure, etc.

10.1 Incineration

It is a thermal process for burning highly combustible waste like plastics, cardboard, paper, and rubber and combustible wastes like cartoons, wood scrap, floor sweepings, and food wastes at a very high temperature. The method does not apply to Indian conditions due to high dust and ash content of wastes; high capital costs, especially for adequate control of emissions; high operation and maintenance costs; and the need for skilled human resources. However, incineration is also associated with the production and release of carcinogenic and toxic compounds. Therefore, the incineration process is not environmentally friendly and is hence usually not recommended as a solid waste disposal technique.

10.2 Pyrolysis

Pyrolysis is also known as thermal pyrolysis. The combustion process is highly exothermic (releasing heat on burning in the presence of oxygen) in nature, whereas pyrolysis is highly endothermic (consuming flame) in the environment. Hence, the process of pyrolysis is known as destructive distillation. In this method, the solid wastes are heated under anaerobic conditions (i.e., burning without oxygen). The organic components of the solid wastes split up into volatile liquid and gaseous fractions (CO, CO_2, CH_4, tar, charred carbon). Pyrolysis cannot handle a wide variety of wastes that exists and will only have a small impact on the overall processing of waste.

11. Problems with existing MSW disposal practices

Transportation is a necessary function for solid waste management activities since municipal solid waste, recyclables, yard waste, and other materials must be collected and transported to be managed. There are various methods for collecting and transporting waste, the choice of which depends on the type of solid waste, the source of solid waste, and the proper management method used.

A vital component of a reliable and well-run solid waste management system is to set up an efficient sanitary landfill and customer-responsive collection and disposal of solid waste. Waste collection services are provided to residents in all cities, either private or self-government agencies. In the meantime, the rapid increase in disposal costs across the city, the cost of collection, and the transfer of wastes continue to raise disposal as a percentage of overall service costs for most communities.

For collection and transfer, waste systems are often complex and challenging to design, because several factors must be considered and a wide range of collection and transfer options are available. The community participation is essential for an efficient MSWM system. However, the municipal authorities have failed to mobilize the community and educate citizens on the principles of handling waste and proper practices of storing it in their bins at the shop, household, etc. Due to the lack of an essential facility of collection of waste from source, citizens are likely to dump waste on the streets, open spaces, drains, and water bodies in the vicinity, creating insanitary conditions. Later, the pourakarmikas will collect the discharged waste through street sweeping, drain cleaning, etc. Street sweeping has thus become the principal method of primary collection.

The tools used for street sweeping are inefficient and outdated. For instance, the broom with a short handle is still in use, forcing sweepers to bend for hours, resulting in fatigue and loss of productivity.

Transportation of waste from the waste storage depots to the disposal site is done through a variety of vehicles, such as three-wheelers, tractors, and trucks. Most of the transport vehicles are old and open. They are usually loaded manually. Due to inadequate workshop facilities and maintenance procedures, the vehicles are poorly maintained. This leads to the breaking down of vehicles, resulting in failure of services for a long time.

The various technological options available for processing, treatment, and disposal of MSW are composting, vermicomposting, AD, incineration, gasification and pyrolysis, production of RDF, and sanitary landfilling.

The main benefits of composting include improvement in soil texture and augmenting of micronutrient deficiencies. It also increases the moisture-holding capacity of the soil and helps in maintaining soil health. However, it is an age-old concept for recycling nutrients to the soil. It does not require significant capital investment compared to other waste treatment options.

The technology of waste-to-energy projects and its viability and sustainability have been proven worldwide. WTE projects involve higher capital investment and are more complicated than other options of waste disposal. These plants are financially viable in developed countries mainly because of the tipping fees charged by the facility for the service of waste disposal, in addition to its revenue income from power sales. However, at present, in Bangalore, revenue from power sales is the only source of income for WTE plants. Technologically, it is feasible to set up even smaller capacity projects of the 1–5 MW range, corresponding to around 100–500 metric tons per day waste treatment. The significant role in making a WTE facility financially viable is the segregation of waste at the source to avoid the mixing of undesirable waste streams.

12. Impact of solid waste on soil and water bodies

MSW landfills are essential in modern-day society because the segregation and disposal of solid waste materials into decentralized locations helps to minimize risks to public health and safety. Currently, in Bengaluru, MSW landfills remain open for decades before undergoing closure and post-closure phases, during which steps are taken to minimize the risk of environmental contamination. Although MSW landfills are an essential part of everyday living, they may present long-term threats to surface water and also hydrologically interlinked groundwater bodies. The impact of leachate on groundwater and surface water bodies has attracted much attention because of its enormous environmental significance. In the olden days, landfills were constructed without leachate collection systems and liners. Once leachate enters the groundwater, it will migrate downward through the unsaturated zone until it finally reaches the saturated area. This resulted in creating significant leachate-contaminate groundwater plumes that follow the hydraulic gradient of the groundwater system.

Unscientific management of MSW leachate will lead to contamination of the soil and water bodies. The presence of a contaminant in the soil can change the engineering properties of the soil. The leachate potential to contaminate the soils and groundwater and surface water bodies assumes significance in the context of existing MSW practices, which have many drawbacks.

Considering the importance of the problem discussed, this research mainly focuses on the characteristics of leachate generated from municipal solid waste landfill sites and its effect on surrounding water bodies near the Mavallipura landfill area in Bengaluru.

13. Reclamation of waste dumps for development in Bengaluru

Many cities, including Bengaluru, are also facing the problems due to old dump yards situated close to the expanding cities. These dump yards need to be reclaimed for the growing needs of the city infrastructure development. This is the case for many cities in India with an alarming rate of the urbanization process. Also, they create a nuisance in the town, and the same needs to be stabilized or reclaimed. The waste sites that were earlier in the periphery of the corporation limits of Bengaluru City are now in the development zone of a more magnificent Bengaluru City, as shown in **Figure 13**. It can be seen from a satellite image that the built-up area has come near and around the earlier dump sites. Thus, it is clear that most of the old dumpsite which is existing around Bengaluru has become potential places for development.

These structures built on these dumpsites can undergo distress due to the high settlement and cause failures due to the low strength of the dumped waste. These dumps can also cause groundwater contamination due to leaching of waste by the percolation of rainwater (**Figures 14–16**).

Naveen [12] carried out a detailed experimental program to study the variations in geotechnical properties for different wastes with a time of dump yard, i.e., in turn with the age of the waste. The data provided on geotechnical properties of the waste with varying degrees of decomposition helps to plan for the reclamation of waste dumps.

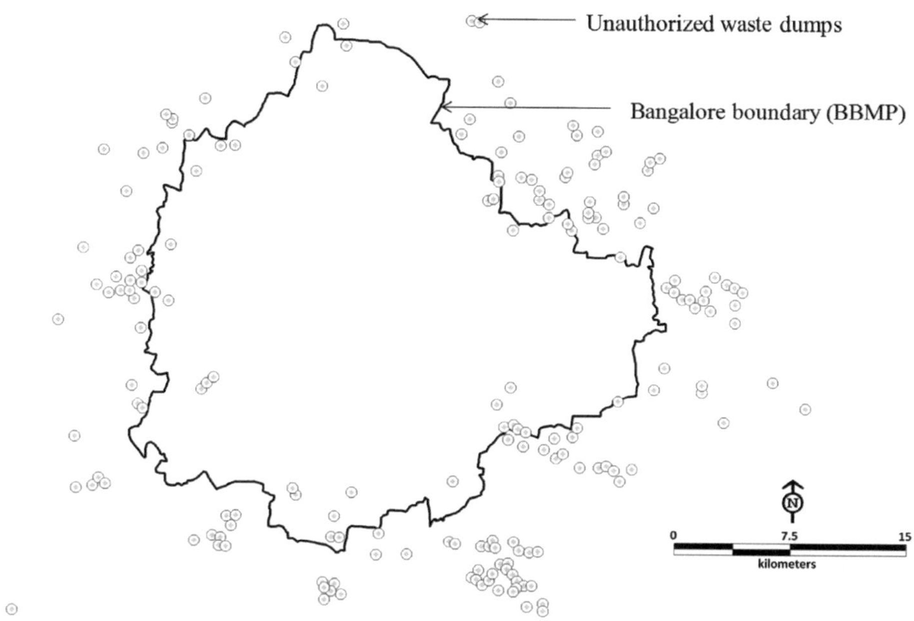

Figure 13.
Location of unauthorized dump sites in and around Bangalore City (Chanakya et al., 2011).

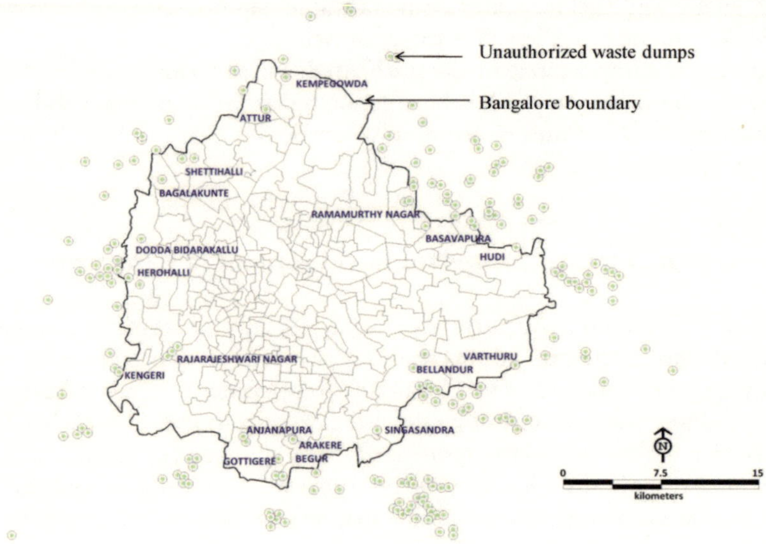

Figure 14.
Unauthorized dumping along with wards in and around Bangalore City.

Figure 15.
Major arterial and sub-arterial road network along with dump sites in and around Bangalore City.

14. Reclamation of MSW landfill

The objective of the reclamation is to return the MSW landfill to a condition as close as possible to leave the site in a state compatible with the surrounding ground. MSW landfill reclamation is a new approach used to expand the MSW landfill capacity and minimize the cost of acquiring additional land. The significant factors influencing the success of reclamation include chemical, hydrologic, and physical conditions of the fill materials, climate, availability of suitable plant species, and proper management of reclaimed sites [13].

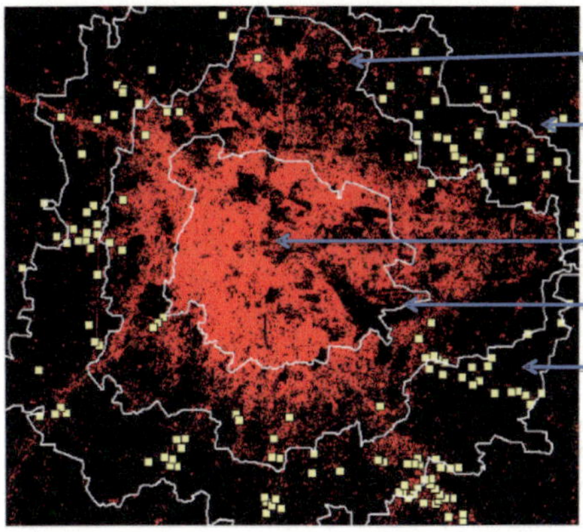

Figure 16.
Satellite image of built-up area along with dump sites in Greater Bangalore.

The essential benefits may include a reduction in closure costs and reclamation of land for other purposes and recovered materials such as recyclables, soil, and waste, which can be burned as fuel. Some drawbacks exist in MSW landfill reclamation. This technology may release methane and other gasses from decomposing solid wastes. Also, the excavation work process involved in reclamation may cause adjacent landfill areas to collapse/sink. Hence, it is necessary to conduct a site characterization study. The site characterization should assess facility aspects, such as geotechnical and geological features and the stability of the surrounding area and identified groundwater, and determine the fractions of good soil, recyclable waste materials, and hazardous waste at the site [14]. Based on the available information from the site characterization, it provides project planners with a basis for assessing the potential economic benefits of a landfill reclamation project.

The economic benefits associated with landfill reclamation are indirect; they may include the following: increased disposal capacity; avoided or reduced cost of landfill closure; revenues from recyclable and reusable materials like ferrous metals, aluminum, plastic, and glass; combustion waste sold as fuel; reclaimed soil used as cover; and land value of sites reclaimed for other uses.

Geotechnical properties of municipal solid waste presume great importance in their reuse, disposal, as well as reclamation of waste and dump sites. Because of the high demand for land, the abandoned, closed landfills have to reclaim to meet the growing needs of the society. Due to several reasons, the population around the improperly operated landfills is demanding the closure of the landfill. However, just leaving the landfill without proper closure cannot be allowed. Thus any attempt to reclaim land for development should come after the characterization of waste for their physical and chemical composition and geotechnical properties. Therefore, these studies constitute the first step to successfully implementing a comprehensive waste management system.

15. Conclusions

Based on the above context, the following conclusions can be drawn:

- For setup, the WTE plants require higher capital investment and are more complicated than the other options of waste disposal.

- WTE plants are suitable in developed countries mainly because of the tipping fees/gate fees charged by the facility for the service of waste disposal, in addition to its revenue income from power sales.
    - Due to the high content of biodegradable waste in Bengaluru, a biological process is needed such as anaerobic digestion and composting to treat the waste, gasification, and pyrolysis.
- Plasma gasification technology can reduce the need for landfills; it can create more renewable energy than the projected energy from solar, wind, landfill gas, and geothermal energies combined.
- RDF plants are in the initial stage of development in India. It is beneficial in preparing an enriched fuel feed for thermal processes like incineration. The RDF pellets are used as a coal substitute at a lower price.
- The pyrolysis process cannot handle the wide variety of wastes, and the end products of pyrolysis are carbon black oil that can be resent to a refiner and hydrocarbon gasses that can be used to make electricity or stream.
- Sanitary landfill is the cheapest, simplest, and most cost-effective method for disposing of waste.

Author details

B.P. Naveen[1*] and P.V. Sivapullaiah[2]

1 Department of Civil Engineering, Amity University Haryana, India

2 CMR University, Bengaluru, India

*Address all correspondence to: bp.naveen864@gmail.com

References

[1] Naveen BP, Sivapullaiah PV, Sitharam TG. Disposal options for solid waste of Bengaluru city based on its characteristics. International Journal of Environment and Waste Management (IJEWM). 2013;**12**(1):77-88

[2] Naveen BP, Sitharam TG, Sivapullaiah PV. Status of solid waste management in Bengaluru and review of solid waste techniques adopted. In: International conference on waste management for sustainable development (21-23 2014); Palakkad, Kerala, India; 2014. pp. 11-17

[3] Dittrich C. Bengaluru: Divided under the impact of globalization. Asian Journal of Water, Environment and Pollution. 2004;**2**(2):23-30

[4] Evaluation of Technology for processing existing waste at Seven Landfill sites of BBMP, Bangalore. Technical Committee Recommendations on EOI Application Report 3 Jan 2013; 2013

[5] Beukering PV, Sehker M, Gerlagh R, Kumar V. Analysing urban solid waste in developing countries: A perspective on Bangalore. 1999. Collaborative Research in the Economics of Environment and Development (CREED) and Environmental Economics Programme (IIED), Working Paper No. 24, March 1999. p. 24. ISBN: 1357 924X (ISSN). Available on the website: http://www.iied.org/eep/ pubs/creed [Accessed: 20 February 2004]

[6] Central Pollution Control Board (CPCB). Collection, Transportation and Disposal of municipal solid waste in Delhi (India)-a case study, CPCB, New Delhi; 2000

[7] BBMP. Solid Waste Management in Bruhat Bangalore Mahanagara Palike, Technical Report; 2007

[8] Ramachandra TV, Bachamanda S. Environmental audit of municipal solid waste management. Technical Report: 118. Bangalore: CES, IISc; 2006

[9] Ministry of Environment and Forests (MoEF). Available from: http://www.envfor.nic.in [Accessed: 16 June 2005]

[10] Dhindaw J, Badami K, Sitharam TG, Chanakya HN. Waste Segregation-a Case Study in Malleshwaram, Bangalore, Technical Report Submitted to CISTUP. Bangalore: IISc; 2014

[11] Sasikumar K, Krishna SG. Solid Waste Management. New Delhi: PHI Learning Private Limited; 2014

[12] Naveen BP. Evaluation of geotechnical properties of municipal solid waste and stabilization of waste dumps [master of science thesis]. IISc, Bangalore; 2012

[13] Naveen BP, Mahapatra DM, Sitharam TG, Sivapullaiah PV, Ramachandra TV. Physico-chemical and biological characterization of urban municipal landfill leachate. Environmental Pollution. 2017;**220** (Part A):1-12

[14] Environmental Protection Agency. Guidelines on the Information to be Contained in Environmental Impact Statements. Wexford: EPA; 2002

Chapter 5

Comparative Life Cycle Assessment of Sewage Sludge (Biosolid) Management Options

Başak Kiliç Taşeli

Abstract

Sludge formation during wastewater treatment is inevitable even with proper management and treatment. However, the proper treatment and disposal of sludge are still difficult in terms of cost of treatment, the presence of new pollutants, health problems, and public acceptance. Conventional disposal methods (e.g., storage, incineration) have raised concerns about legislative constraints and community perception that encourage the assessment of substitute sludge management options. Sludge management requires a systematic solution that combines environmental effectiveness, social acceptability, and economic affordability. Life cycle assessment is one of the most important tools to identify and compare the environmental impact of sludge treatment technologies to ensure sustainable sludge management. Increased production of sludge (biosolids) increases worldwide due to population growth, urban planning, and industrial developments. The sludge needs to be properly treated and environmentally managed to reduce the negative effects of its application or disposal. This chapter deals with the application of biosolids or sewage sludge, together with possible resources for sustainable development. In this section, the life cycle assessments of sludge treatment methods were also investigated and found that sludge treatment techniques lead to major environmental impact categories such as global warming potential, human toxicity, acidification potential, and resource consumption.

Keywords: life cycle assessment, sludge management, biosolid management, sustainability, sludge treatment

1. Introduction

Wastewater treatment plants (WWTP) are aimed to decrease the environmental impacts of discharging untreated wastewater into receiving bodies, but considering the need for long-term ecological sustainability, the objectives of wastewater treatment systems should include energy and resource savings and waste reduction [1].

Sewage sludge management is a management system that makes sludge recovery a central component of a wastewater treatment plant that strives to integrate it with improving the sustainability of wastewater plants. Currently, wastewater sludge production, treatment, and disposal methods vary from country to country, and the continued growth of sludge production is becoming a global problem. The sludge production rate is increasing due to the stricter legislation which is constantly solidifying for the sustainable disposal of wastewater. Nowadays, however, due to the increasing environmental awareness of the public and increasing pressure from environmental organizations, sludge management has become necessary with economic and environmentally friendly methods.

Conventional disposal methods (e.g., landfill, incineration, stabilization) have raised concerns about legislative constraints and community perception that encourage the assessment of substitute sludge management options. Sludge management requires a systematic solution combining environmental effectiveness, social acceptability, and economic affordability based on a life cycle approach. Life cycle assessment is one of the most important tools to identify and compare the environmental impact of sludge treatment technologies to ensure sustainable sludge management.

Generally, the terms biosolids and sewage sludge are used interchangeably. Biosolid includes 20% content of fat, 50% carbohydrate content, 30–40% content of organic matter, 3% total nitrogen, 1.5% total phosphorus, 0.7% total potassium content, 10–20% C/N ratio, pH of 6.5–7.0, and a specific gravity of 1.00 as reported by [2]. It is a by-product of treatment plants in large quantities varying in characteristics, containing organic and inorganic chemicals, heavy metals (iron, chromium, manganese, zinc, mercury, lead, nickel, cadmium and copper), and pathogens. It is considered as a resource due to the widespread application in biogas production, soil filling, organic fertilizer, and soil amendment.

Life cycle assessment (LCA) is a standardized and recognized tool to measure the overall environmental impact of providing a product or service. It is increasingly used to support commercial claims of products' environmental performance. It is also used as the basis for European environmental legislation, including Integrated Pollution Prevention and Control (IPPC) and Integrated Product Policy [3].

ORWARE, SimaPro, MARTES, UMBERTO, Ecobilan, LCAiT, SiSOSTAQUA, BioWin*, STAN*, GaBi 6, WWEST, BEAM, GEMIS, and Quantis Suite are the best known commercial sludge treatment and management LCA software. Among them SimaPro is the most widely used model. Global warming, acidification, eutrophication, photochemical smog, human toxicity, ecotoxicity, depletion of abiotic resource, and terrestrial ecotoxicity are the fundamental impact categories of sludge management [4–12]. Life cycle assessment methodology is generally implemented for the main sludge management like dewatering, thickening, and anaerobic digestion [12–18].

This chapter deals with the application of biosolids or sewage sludge, together with possible resources for sustainable development. Moreover, the life cycle assessments of sludge treatment methods were also investigated and found that sludge treatment techniques lead to major environmental impact categories such as global warming potential, human toxicity, acidification potential, and resource consumption.

2. Life cycle assessment (LCA)

Life cycle analysis (LCA) is a method of assessing the environmental impact of products and processes throughout their lives, including raw material procurement, production, use, final disposal, and all transport phases between these stages.

With this analysis, the comprehensive inventory of all energy, water, and substance inputs together with the emitted waste is evaluated together, and the possible environmental effects of the products are calculated. Unlike other narrow-scale environmental impact analyses, the LCA examines environmental issues with its "cradle-to-grave" approach [19].

Depending on the stage of the life cycle, LCA studies can be grouped as "cradle to grave," "cradle to door," "cradle to cradle," and "door to door." It is a definition used for studies examining the whole life cycles of a product or process from cradle to grave and includes all the processes to be passed from the raw material production (cradle) to the disposal of the waste (grave).

A product or process from the cradle to the door partially covers the processes from the production of the raw material (cradle) to the stage (door) to which it is delivered to the factory. This is a life cycle study which partially covers processes.

The recycling of waste during the waste disposal phase is referred to as the cradle-to-cradle approach.

"Door to door" is an approach that deals with the life cycle of a single stage of a product or a process [20].

LCA is a rapidly evolving tool designed to help environmental management in sustainable products and services in the longer term, also called "life cycle analysis," "life cycle approach," "cradle-to-grave analysis," or "ecological balance."

The standard LCA method consists of four main steps:

First step: aim and scope definition: At this stage, the objective of LCA study, target groups, basic variables, necessary data, constraints, and assumptions used are defined. Systematic and functional units are the two most important elements defining the scope and knowledge of the study, and while determining the system boundaries, the life cycle of the product is included in the analysis [21]. The functional unit refers to the unit function of the system under consideration and should be expressed clearly and in detail and should reflect the basic function of the product or system [22].

Second step: life cycle inventory analysis: At this stage, energy, water, raw material inputs, and released solid waste, wastewater, and air emissions are determined within the boundaries. In the meantime, inventory information about all unit processes in the product's life cycle is compiled through data collection forms, and deficiencies are completed by using literature review and sectoral reports.

All collected data is rearranged according to the functional unit. It is made available for the calculation of environmental impacts. At this stage, data quality and accuracy are vital at every step.

As a result of the literature research, it was determined that UMBERTO, GEMIS, SimaPro, GLOSSARY BEAM, MARTES, Ecobilan, LCAiT, SiSOSTAQUA, BioWin, STAN, GaBi 6, and WWEST are the most preferred sludge treatment and management LCA software [4–12]. Among them SimaPro is the most widely used model. Global warming, acidification, eutrophication, photochemical smog, human toxicity, ecotoxicity, depletion of abiotic resource, and terrestrial ecotoxicity are the fundamental impact categories of sludge management. Life cycle assessment methodology is generally implemented for main sludge management like dewatering, thickening, and anaerobic digestion [12–18].

Third step: life cycle impact analysis: At this stage, environmental impact potentials are calculated using inventory data collected and compiled in the previous stage. Mandatory (classification and characterization) and voluntary (normalization and weighting) substages of the impact analysis stage are defined in [23, 24]:

a. At the classification stage, the individual inventory items are assigned according to the relevant environmental impact categories. For example, CO_2 emissions are categorized as "global warming." The most commonly used environmental impact categories in LCA studies are acidification, eutrophication, global warming, photochemical ozone formation, ozone depletion, ecotoxicity, and resource consumption (see **Figure 1**).

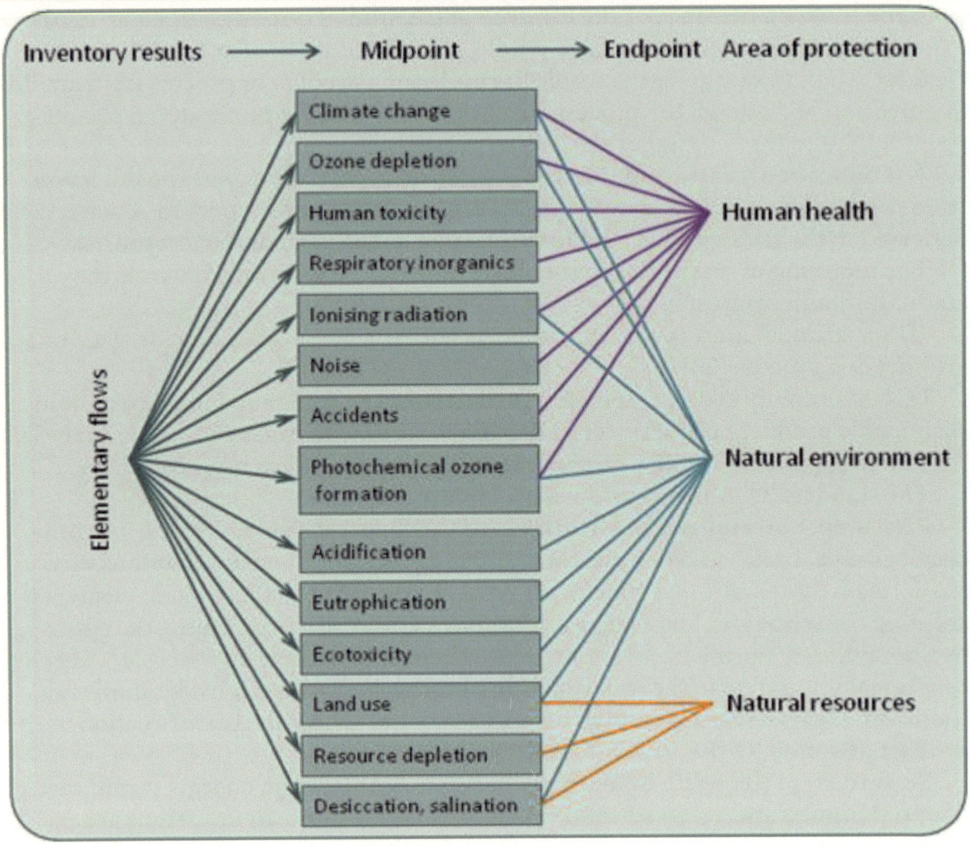

Figure 1.
LCA impact assessment mechanism [19].

b. In the characterization stage, inventory items contributing to the same environmental problem are multiplied by the relevant coefficients and expressed over the common unit, and the aggregated impact is calculated for each environmental impact category. For example, CO_2, CH_4, and N_2O emissions leading to global warming are expressed by an equivalent of kg CO_2.

c. In the normalization phase, different environmental impact potentials are compared according to the common reference system using accepted normalization methods. Normalization indicates which environmental impact potential is higher.

d. In the weighting phase, the normalization results are multiplied by coefficients using one of the weighted methods that are accepted and based on the reduction targets for each environmental impact category. Weighting reveals which environmental impact potential is more important. LCA has a wide range of applications in the private, public, and academy sector for a wide range of products, services, and systems. LCA develops strategic planning, public policies, and performance indicators; identifies priority products and processes in production; identifies improvement opportunities; provides important inputs in product development or redesign stages, various sustainability declarations, and eco-label programs; and support and compare different production alternatives. Among these, environmental declarations and carbon and water footprint calculations are important in sustainable consumption and production.

Fourth step: interpretation of results: It is the purpose of this stage to interpret the results of the inventory and environmental impact analysis stages according to the purpose and scope of the study and to present important results and recommendations for the system or product under consideration [23, 24].

The carbon footprint is an environmental indicator that measures the global climate change caused by greenhouse gases (GHG) from the life cycle of products and services. Greenhouse gases emitted at each stage are expressed in terms of the total CO_2 equivalent units multiplied by the relevant coefficients. Basically carbon footprint calculation is also a life cycle approach; however, unlike LCA, it does not cover all emissions but only inputs that contribute to global warming. Principles and procedures for product carbon footprint calculation and declaration are defined in [24], ISO 14067: 2013. Here, ISO 14040 and ISO 14044 LCA standards are used for footprint calculations, and ISO 14020, ISO 14024, and ISO 14025 standards are used for declarations [24].

Water footprint calculations are likewise based on life cycle principles, and the basic rules and principles for footprint calculations of products, processes, and institutions are included in the ISO 14046: 2014 standard. With this method, in addition to direct and hence water input-outputs, air and soil emissions affecting water quality are also addressed [25].

The main uses of LCA can be summarized as follows:

a. Analyzing problems related to a specific product

b. Determining the important parameters that affect a study for product development

c. New product design

d. Choosing between similar products, processes, and services

One of the areas where LCA is used in particular is green purchasing applications. Eco-labels (where environmentally friendly products are documented) are preferred by consumers; Blue Angel is used in Germany, but in Scandinavian countries Green Swan is used.

Below are examples of other uses:

a. Compliance of the various packaging alternatives with the European Union's packaging directive

b. Evaluating the different waste management approaches of the municipalities

c. Comparing different types of biomass for a particular use (e.g., obtaining electricity) to determine environmental advantages and disadvantages

d. Strategic comparison between alternatives in order to make a decision on a public investment, for example, evaluation of transport methods (road, rail, sea) for certain regions or a particular sector

e. Harmonization of the construction sector with the environment

f. Improving the raw material production stage by switching to sustainable raw materials in production

g. Reducing the carbon footprint by increasing the energy efficiency of the electronic goods produced through R&D studies

h. Making product shipments more efficient and reducing air emissions by making changes in product packaging

i. Reducing the environmental impact of the final disposal phase by designing more recyclable products

As a summary the LCA study, which covers all stages of the product value chain, evaluates the total environmental impact such as global warming, acidification, eutrophication, photochemical smog, human toxicity, ecotoxicity, depletion of abiotic resource, and terrestrial ecotoxicity which are the fundamental impact categories of sludge management [26].

3. Life cycle assessment of sewage management

A sludge management that yields the best results requires a systematic solution that combines environmental effectiveness, social acceptability, and economic affordability based on a life cycle approach. For example, it is reported in the literature that total sludge production in China increased by an average of 13% per annum from 2007 to 2013, producing 6.25 million tons of dry solids in 2013 and reaching 39.78 million tons in 2020. In the same study, more than 80% of the sludge was disposed without any process, the organic content of the sludge was around 37%, and because of this low organic matter content, anaerobic digestion was not an efficient method, and therefore storage and incineration after dewatering was the most common method [27]. With the increase in population, the urban settlement areas expanded, and many of the wastewater treatment plants in the rural areas are now located within the settlements, and the gradual improvement in socioeconomic development and living standards has led the residents to pay more attention to the quality of the living environment [1]. In addition, since the environmental awareness of the public has increased, the odors caused by treatment plants have become a new and troublesome social problem. Another study suggests that municipal waste management in China tends to incinerate instead of landfill, but it causes social conflicts as it impedes the construction of treatment plants near public land or habitats [28].

Another study reached a more striking conclusion, stating that the inhabitants of 12 Chinese cities protested the incineration projects due to environmental concerns and only three incineration projects were allowed from 2009 to 2015, while others were canceled and the disposal of sludge into the long-term storage or incineration facilities was increasingly difficult [29].

Since EU member states have to reduce the amount of biodegradable municipal waste to 35% of 1995 by 2016, they have to make the transition from a linear to circular economy where waste can be converted into resources. Therefore, like all biodegradable wastes, wastewater sludge is seen as a source of energy and material production. However, a sewage management should be considered, including the method of processing sludge, where and how the final products (e.g., fertilizer, biogas) are used, the amount of greenhouse gas (GHG) emissions, and the selection of the most sustainable wastewater sludge treatment technology. Composting, anaerobic digestion, incineration, chemical stabilization, and the use of fertilizer in agricultural land are the most commonly used sewage management methods. The change of sludge management methods from country to country was mentioned in the previous chapters. Portugal, Ireland, the United Kingdom, Luxembourg, and Spain use more than 75% of the sludge produced as fertilizer for agricultural land; 86% of the sludge produced in Lithuania, Finland, and Estonia is composted; the Netherlands, Belgium, Germany, Slovenia, Austria, and Switzerland prioritize incineration, while Malta, Romania, Italy, and Bosnia and Herzegovina mostly report the use of sludge for storage [30].

In a sludge life cycle assessment study in France, it was noted that the final combination of anaerobic digestion and land application caused the lowest emissions during operation [5]. Many researchers have indicated that, from an economic point of view only, a large-scale incineration plant or anaerobic digester may be the most effective way to treat sludge [30].

It was reported that land filling has the greatest impact (296.9 kg CO_2 eq./t sludge), followed by mono-incineration (232.2 kg CO_2 eq./t sludge) and carbonization (146.1 kg CO_2 eq./t sludge) in terms of the emission quantity of greenhouse gases. They also stated that co-incineration with municipal solid waste has the benefit of reducing greenhouse gas emission by −15.4 kg CO_2 eq./t sludge [31].

A calculator calculating the greenhouse gas (GHG) emissions (carbon dioxide including bio-based, methane, and nitrous oxide measured as carbon dioxide equivalents) from sewage sludge treatment methods found at the end of the comprehensive study showed that composting, anaerobic digestion, and incineration resulted in the lowest emissions of the GHG gases. If you need to elaborate further, anaerobic digestion generated the least carbon dioxide equivalent emissions among all the treatment methods studied. The second best option was incineration of sludge, while the third best was composting [31].

In another study a life cycle assessment (LCA) was performed on five common sewage sludge treatment practices, namely, dewatering of mixed sludge, lime stabilization of dewatered sludge, anaerobic digestion of mixed sludge, dewatering of anaerobically digested sludge, and incineration of dewatered anaerobically digested sludge. The sludge residues were applied on agricultural land, and it was found that the incineration of dewatered anaerobically digested sludge scenario performed better results [2].

Ten impact categories, namely, human toxicity carcinogenic effects, human toxicity non-carcinogenic effects, ecotoxicity, freshwater eutrophication, marine eutrophication, terrestrial eutrophication, terrestrial acidification, particulate matter formation, climate change, and photochemical oxidant formation, were also assessed in this study. It was concluded that in human toxicity and ecotoxicity categories, impacts were dominated by the application of zinc and copper to agricultural soil. For the freshwater eutrophication potential category, the fate of phosphorus was found to be (P) dominated, while the fate of N had a profound effect on all nontoxic impact categories other than freshwater eutrophication [2].

4. Conclusions

As a result of the literature blended in this section, it is concluded that biosolids have significant disadvantages for their use in agriculture and other applications and, therefore, sludge or biosolids should be sampled, controlled, and monitored regularly for pollutants (pathogens, heavy metals, etc.). However, it is also concluded that biosolids play an important role in energy production, and crop production. The most comprehensive sludge management studies have shown that land application is an important contribution to global warming, eutrophication, and acidification. More scientific research is needed on different aspects of biosolids or sewage sludge to be a more suitable resource for sustainable development. It is vital that the most efficient sludge management strategy should focus on economic, technological, and societal constraints.

The LCA study, which covers all stages of the product value chain, evaluates the total environmental impact such as global warming, acidification, eutrophication, photochemical smog, human toxicity, ecotoxicity, depletion of abiotic resource, and terrestrial ecotoxicity which are the fundamental impact categories of sludge management.

A literature blending in GHG showed that composting, anaerobic digestion, and incineration have the lowest emissions. Many researchers have indicated that, from an economic point of view only, a large-scale incineration plant or anaerobic digester may be the most effective way to treat sludge. A sludge management that yields the best results requires a systematic solution that combines environmental effectiveness, social acceptability, and economic affordability based on a life cycle approach.

Author details

Başak Kiliç Taşeli
Environmental Engineering Department, Faculty of Engineering, Giresun University, Giresun, Turkey

*Address all correspondence to: basak.taseli@giresun.edu.tr

References

[1] Xiao L, Lin T, Wang Y, Ye Z, Liao J. Comparative life cycle assessment of sludge management: A case study of Xiamen. China. Journal of Cleaner Production. 2018;**192**:354-363

[2] Kumar V, Chopra AK, Kumar A. A review on sewage sludge (biosolids) a resource for sustainable agriculture. Archives of Agriculture and Environmental Science. 2017;**2**(4):340-347

[3] Available from: https://www.eiga.eu/index.php?eID=dumpFile&t=f&f=251&token =6df78a58d 435b481e0c 692d7a2ba67cdbe9deb0b [Accessed: 26 November 2019]

[4] Wiloso EI, Heijungs R, Snoo GR. LCA of second generation bioethanol: A review and some issues to be resolved for good LCA practice. Renewable and Sustainable Energy Reviews. 2012;**16**:5295-5308

[5] Suh YJ, Rousseaux P. An LCA of alternative wastewater sludge treatment scenarios. Resources, Conservation and Recycling. 2002;**35**:191-200

[6] Gallego A, Hospido A, Moreira MT. Environmental performance of wastewater treatment plants for small populations. Resources, Conservation and Recycling. 2008;**52**:931-940

[7] Murray A, Horvath A, Nelson KL. Hybrid life-cycle environmental and cost inventory of sewage sludge treatment and end-use scenarios: A case study from China. Environmental Science and Technologies. 2008;**42**: 3163-3169

[8] Lederer J, Rechberger H. Comparative goal-oriented assessment of conventional and alternative sewage sludge treatment options. Waste Management. 2010;**30**:1043-1056

[9] Sablayrolles C, Gabrielle B, Montrejaud VM. Life cycle assessment of biosolids land application and evaluation of the factors impact in human toxicity through plant uptake. Journal of Industrial Ecology. 2010;**14**:231-241

[10] Hospido A, Carballa M, Moreira M. Environmental assessment of anaerobically digested sludge reuse in agriculture: Potential impacts of emerging micropollutants. Water Research. 2010;**44**:3225-3233

[11] Bravo L, Ferrer I. Life cycle assessment of an intensive sewage treatment plant in Barcelona (Spain) with focus on energy aspects. Water Science and Technology. 2011;**64**(2):440-447

[12] Linderholm K, Tillman AM, Mattsson JE. Life cycle assessment of phosphorus alternative for Swedish agriculture. Resources, Conservation and Recycling. 2012;**66**:27-39

[13] Tillman AM, Svngby M, Lundstrom H. Life cycle assessment of municipal waste water systems. International Journal of Life Cycle Assessment. 1998;**3**:145-157

[14] Bridle T, Skrypski-Mantle S. Assessment of sludge reuse options: A life-cycle approach. Water Science and Technology. 2000;**41**:131-135

[15] Lundin M, Bengtsson M, Molander S. Life cycle assessment of wastewater systems: Influence of system boundaries and scale on calculated environmental loads. Environmental Science and Technology. 2000;**34**:180-186

[16] Soda S, Iwai Y, Sei K. Model analysis of energy consumption and greenhouse gas emissions of sewage sludge treatment system with different

processes and scales. Water Science and Technology. 2010;**61**(2):365-373

[17] Brown S, Beecher N, Carpenter A. Calculator tool for determining greenhouse gas emissions for biosolid processing and end use. Environmental Science and Technology. 2010;**44**:9509-9515

[18] Nakakubo T, Tokai A, Ohno K. Comparative assessment of technological systems for recycling sludge and food waste at greenhouse gas emissions reduction and phosphorus recovery. Journal of Cleaner Production. 2012;**32**:157-172

[19] EPA, U.S. Environmental Protection Agency. Life Cycle Assessment: Principles and Practice. Washington, DC. EPA/600/R-06/060; 2006

[20] Jiménez-González C, Kim S, Overcash M. Methodology for developing gate-to-gate life cycle inventory information. The International Journal of Life Cycle Assessment. 2000;**5**:153-159

[21] Tillman AM, Ekvall T, Baumann H, Rydberg T. Choice of system boundaries in life cycle assessment. Journal of Cleaner Production. 1993;**2**:21-29

[22] Wiloso EI, Heijungs R, Snoo GR. LCA of second generation bioethanol: A review and some issues to be resolved for good LCA practice. Renewable and Sustainable Energy Reviews. 2012;**16**:5295-5308

[23] ISO: Environmental management– Life cycle assessment: Principles and framework. ISO 14040:2006, ISO, Geneva, Switzerland; 2006

[24] ISO: Greenhouse gases .Carbon footprint of products: Requirements and guidelines for quantification and communication. ISO/TS 14067:2013, ISO, Geneva, Switzerland; 2013

[25] ISO: Environmental management. Water footprint: Principles, requirements and guidelines. ISO 14046:2014, ISO, Geneva, Switzerland; 2014

[26] Yilmaz O, Anctil A, Karanfil T. LCA as a decision support tool for evaluation of best available techniques (BATs) for cleaner production of iron casting. Journal of Cleaner Production. 2015;**105**:337-347

[27] Yang G, Zhang G, Yang H. Current state of sludge production, management, treatment and disposal in China. Water Research. 2015;**78**:60-73

[28] Xiao LS, Wang R, Chiang PC. Comparative life cycle assessment (LCA) of accelerated carbonation processes using steelmaking slag for CO2 fixation. Aerosol and Air Quality Research. 2014;**14**(3):892-904

[29] Deaton BJ, Hoehn JP. Hedonic analysis of hazardous waste sites in the presence of other urban disamenities. Environmental Science and Policy. 2004;**7**(6):499-508

[30] Piippo S, Lauronen M, Postila H. Greenhouse gas emissions from different sewage sludge treatment methods in north. Journal of Cleaner Production. 2018;**177**:483-492

[31] Wang N-Y, Shih C-H, Chiueh P-T, Huang Y-F. Environmental effects of sewage sludge carbonization and other treatment alternatives. Energies. 2013;**6**:871-883

Section 2

Green Economy

Chapter 6

Circular Economy and Green Public Procurement in the European Union

Jarosław Górecki

Abstract

Until now, construction was considered through the prism of technical possibilities of implementing investment plan supporting and, at the same time, urbanization processes. The development model, present in highly developed countries, is far from sustainable. The departure from natural technologies for erecting construction works must have resulted in excessive use of resources, mainly nonrenewable. The strong negative impact on the natural environment of the architecture, engineering, and construction (AEC) industry cannot go unnoticed. Therefore, a solution to the problem of excessive energy consumption in technological processes in construction, which are also generators of huge amounts of pollution, should be discovered. Circular economy (CE) is one of the concepts of response to the threat posed by these negative externalities. It is worth considering construction materials as reusable elements, e.g., after the demolition of a building. The implementation of the CE concept in AEC requires an identification of the next stage in the life cycle of buildings—the rebirth. The chapter focuses on the issues of green public procurement present in the orbit of interest of decision-makers from the European Union. It was associated with the idea of CE, which is significantly entering the construction sector in both managerial and technical terms.

Keywords: circular economy, green public procurement, construction, processes, investment

1. Introduction

The first, and probably one of the most important steps in construction projects, from which the organization of construction works starts, impacting the entire course of the project life, is to get all appropriate building materials.

Imagine the following situation. The basic raw materials used to erect buildings include stone, brick, lime, sand, and wood. Brick, as an innovative material, is becoming more and more popular. It begins to displace wood, which until now has been the main material used in constructions. Even though new ceramic technology is developing rapidly and successfully, not all investors use it. The poorer rural areas are still dotted here and there with thatched cottages. The use of stone remains wide, usually for the needs of foundations. Lime and sand are components of binders without which it would be impossible to permanently connect separated elements of the structure, called semifinished products. It is worth noting that feudalism effectively limits the development of quarries and brickyards. However, the effective transport of purchased goods depends mainly on the distance from the factory to the built-in location. Transportation of building materials is extremely expensive and requires the provision of a sufficiently high number of means of transport, i.e., horse-drawn or oxen-drawn wagons. Transport accounts for a significant percentage of construction costs and sometimes equals or even exceeds the value of transported materials. These reasons cause that the construction industry suffers from a permanent shortage of materials. Insufficient production capacity can be evidenced by the common recovery of demolition building materials [1]. This process is an alternative to the linear production model, in which the deficiencies described earlier effectively limit the development of societies.

This could be a perfect genesis of the idea of circular economy (CE) in the construction sector. The realities presented in the source texts dating back to the Middle Ages are close to those present in the twenty-first century. Unfortunately, in the meantime, there have been some twists and turns that on the one hand effectively limit thinking about construction as an eco-friendly industry and on the other hand that there is no turning back from radical moves and changes.

One of the turning points was the successful research on polymers carried out in the twentieth century. Since the 1950s, it is the moment when mass production of plastics began; over 8000 million metric tons (Mt) were produced in total [2]. The lion's share of this production goes to the construction industry [3] in the form of materials and packaging. Their advantages often overcome the disadvantages that are unacceptable from an ecological point of view. Synthetic materials disintegrate for a very long time, and from the point of view of even several consecutive generations, a majority of them are practically not degradable. Globally, the majority of plastic waste goes to landfills, not always legal ones, and from there to the seas and oceans. The increase in pollution caused by the presence of plastic in the water is frightening [4–6]. The augmented mortality of marine life (fish, marine mammals, flora), as well as the potential threat of the presence of microplastics in the food chain (of which human being is a part), caused that the problem really begins to be discussed. Political decisions are inevitable, but personal habits require a drastic, immediate change.

Households are subject to some consistent waste management policies in many countries. Unfortunately, construction sites are not restrictively treated as, e.g., individual properties, there are not so many fractions, and the garbage received is often mixed and unsuitable for reuse or further processing. But negative externalities of the construction industry are not just solid wastes. There are other pollutants and emissions generated throughout the entire life cycle of construction projects. The problem has been increasing step by step.

That is why the European Union bodies decide to significantly change its legal regulations or to create new guidelines which are focused on encouraging authorities and individual people to return to sustainable development. Since 2010, the European Commission has been sharing lessons learned on green public procurement (GPP) to show how public authorities in the European Union have successfully "greened" public tenders and procurement processes. GPP was defined in the Communication entitled "Public procurement for a better environment" as "a process whereby public authorities seek to procure goods, services and works with a reduced environmental impact throughout their life cycle when compared to goods, services and works with the same primary function that would otherwise be procured" [7].

This chapter concentrates on the architecture, engineering, and construction (AEC) industry and its impacts on the environment. All issues related to the concept of circular economy and green public procurement were shown in the light of this sector which is treated as the most significant source of contaminants. The research covers a literature review on the CE concept and GPP. The results of the study on the ecological quality of construction processes were included too. Besides, a contribution of the chapter is to show a proposal of the eco-friendly vision of AEC supported by CE-based procedures implemented in GPP strategy in the European Union.

2. Circular economy concept

Many concepts limiting a negative impact on the environment are nowadays promoted all over the world. Circular economy has become a solution that theoretically provides significant relief to nature. To make this concept not just a substitute for a somewhat diminished "sustainable development," it is expected that radical changes in shaping natural resource management policies are created.

According to Ellen MacArthur Foundation, a famous worldwide trendsetter of the concept of a circular economy, a transition from a linear model of production to closed-loop variant helps to work effectively at all scales [8]. It does not cover only some adjustments aimed at reducing the negative externalities of the traditional economic paradigm. It simply represents a systemic transition that builds long-term resilience and provides environmental and social reliefs.

AEC, as a sector with high resource consumption, is a good example for explaining how far CE may be useful. It is one of the world's largest waste generators [9]. At the same time, it consumes 40% of the materials entering the global economy and generates 40–50% of the global output of greenhouse gas emissions [10]. Therefore, this sector cannot be considered as environmentally friendly. However, due to recent observations, even in the AEC sector, decision-makers are wondering how to implement some radical changes aiming to reverse the fate of the impending environmental disaster.

The Ellen MacArthur Foundation underlines that the term of circularity has a deep historical and philosophical background. However, with current advances, information technology has the power to support the transition to a circular economy by radically increasing virtualization, transparency, and feedback-driven intelligence. CE model promotes the notion to make more sustainable production models, which are based on careful management of resources and the reduction of negative impacts. Its applications can foster significant improvements in the sustainability of the AEC sector.

There are different perspectives for analyzing the problem of circular economy in the construction sector: from technological issues, to the constructability of the solutions based on the zero-waste attitude and management perspective (only what gets measured gets done [11]), to system problems concerning the whole life cycle of the projects [12] and strategic perspective. In addition, it has to be said that planning the colonization of space requires solid rudiments. It seems that CE can be also applicable to such long-range plans of humanity.

Scientists are building the theoretical rudiments for the new concept [13]. New CE-related professions emerge. Therefore, proper preparation for such a revolution is needed. The methods of selecting suitable candidates for the position of circular economy manager were developed [14]. Systemic changes are also needed [15, 16].

The following concepts like biomimicry [17], industrial ecology [18], cradle to cradle [19], and design for deconstruction [20] are inseparably connected with the concept of CE in the AEC sector.

It turns out that CE is becoming an exemplary attitude for decision-makers when it comes to public procurement.

3. Green public procurement (GPP)

For almost 10 years, the European Commission has been promoting a voluntary instrument connected with good practice experiences on green public procurement. It helps to illustrate how public authorities all over Europe have successfully "greened" a public tender/procurement process. There are many ideas, methods, and tools to expand environmentally friendly attitudes towards business and public development. Among others, they are circular economy concept, sustainable innovations, life cycle costing, etc. Therefore, GPP can be treated as a strategy in which public institutions try to obtain goods, services, and works whose environmental impact during their whole life cycle is smaller than other variants of identical purpose that would be ordered otherwise. It tries to encourage market players to convert their ways of thinking into more sustainable. It attracts decision-makers' interest in the possible alternatives in terms of making the best offer selection more effective. As a part of the new solution, there are good practice cases published online [21], accessible to all interested parties, which provide some suggestions for replicating experiences. There are 22 sections, ordered alphabetically, where one can find different case studies described carefully and focused on making procurement processes less harmful to plants, animals, and other organisms that live on Earth.

According to the European Commission [22], green public procurement can provide public authorities with financial savings. Taking into account the cost of ordered products or services throughout their life cycle can reveal that a selection based only on the price of the purchase can mislead the decision-makers and encourage them to choose not the best offer. However, an awareness of public authorities is rather low. While GPP stays a voluntary procedure, it is important to educate people responsible for procurement processes and explain to them what really pays off. For example, buying products with low-energy or water consumption can lead to a significant reduction in utility bills. Lowering the share of hazardous substances in purchased products (goods or services) can limit the cost of disposal or recycling. Moreover, the bodies responsible for the GPP implementation will be prepared to meet changing environmental challenges as well as to achieve targets for reducing CO_2 emissions and increasing the energy efficiency of products manufactured in the European Union.

3.1 Legal regulations

Each EU member state has to follow some legal regulations. There are basically three areas in the field of legislation related to green public procurement: national law, EU law, and other laws. As for national law, the member states introduce laws together with a number of regulations as implementing acts to those legal acts that specify the nature of public procurement proceedings. Their content is adapted to promote GPP. Then, there is the EU law, which is conditioned by the Directive 2014/24/EU of the European Parliament and of the Council of 26 February 2014 on public procurement and repealing Directive 2004/18/EC [23]. It addresses environmental issues in the following areas:

- Award criteria
- Contract performance conditions
- Environmental management standards

- Grounds for exclusion
- Labels
- Life cycle costing
- Qualification criteria
- Technical specifications

Other laws are formal records related to GPP but not necessarily connected with the core of procurement matter. These are:

- Regulation (EU) 2017/1369 of the European Parliament and of the Council of 4 July 2017 setting a framework for energy labelling and repealing Directive 2010/30/EU
- Directive 2012/27/EU of the European Parliament and of the Council of 25 October 2012 on energy efficiency, amending Directives 2009/125/EC and 2010/30/EU and repealing Directives 2004/8/EC and 2006/32/EC
- Directive 2010/31/EU of the European Parliament and of the Council of 19 May 2010 on the energy performance of buildings
- Regulation (EC) No. 66/2010 of the European Parliament and of the Council of 25 November 2009 on the EU Ecolabel
- Regulation (EC) No. 1221/2009 of the European Parliament and of the Council of 25 November 2009 on the voluntary participation by organizations in a community eco-management and audit scheme (EMAS), repealing Regulation (EC) No 761/2001 and Commission Decisions 2001/681/EC and 2006/193/EC
- Regulation (EC) No. 1222/2009 of the European Parliament and of the Council of 25 November 2009 on the labelling of tires with respect to fuel efficiency and other essential parameters
- Directive 2009/33/EC of the European Parliament and of the Council of 23 April 2009 on the promotion of clean and energy-efficient road transport vehicles
- Regulation (EC) No. 106/2008 of the European Parliament and of the Council of 15 January 2008 on a Community energy efficiency labelling program for office equipment

3.2 Environmental criteria

The European Commission has taken some steps to create common criteria for GPP that can be used in all EU member states. They were developed for those product groups that were considered as the most suitable for GPP implementation. The criteria are the result of close cooperation between the services of the European Commission and other stakeholders. An application of the criteria is nonobligatory. They were formulated so that, after some minor changes, they could be included (partly or fully) in the procurement documentation by a body. In the AEC sector, the most relevant criteria are for:

- Sanitary tapware [24] (last update, 2013)
- Toilets and urinals [25] (last update, 2013)
- Waste water infrastructure [26] (last update, 2013)
- Water-based heaters [27] (last update, 2014)
- Road design, construction and maintenance [28] (last update, 2016)
- Office building design, construction and management [29] (last update, 2016)
- Paints, varnishes and road marking [30] (last update, 2018)
- Road lighting and traffic signals [31] (last update, 2018)
- Road transport [32] (last update, 2019)
- Public space maintenance [33] (last update, 2019)

All the above requirements generally aim to find a balance between environmental performance, economic effectiveness, market availability, and controlling accessibility.

In order to understand the European development model based on GPP and CE, theoretical considerations on ecology should be presented.

4. Ecological engineering vs. theory of ecology

Practical applications of the theory of ecology are connected with a scope of ecological engineering. This phenomenon can be understood as a field of applied sciences, which is the basis for rational use and protection of the environment as well as natural and anthropogenic resources. It can be described as a design of sustainable ecosystems that integrate human society with its natural environment for the benefit of both [34]. Being the nexus of ecology and engineering design, ecological engineering is a distinct engineering discipline [35]. It is used for the ecological development of societies. Ecological engineering deals with the development of new procedures in case where the classical ones are based on assumptions that cannot be real. At the same time, it is based on theoretical knowledge in the field of the general theory of ecology. Ecological engineering solutions also generate issues for general considerations, developing the theory of ecology covering life and technical science, economy, and social science. A complementarity of engineering and ecology theory is presented in **Figure 1**.

On the other hand, according to Allen et al. [36], environmental engineering is an extension of the engineering process that considers the environment in as many aspects as are thought to be relevant. Environmental engineering, as opposed to ecological engineering, works only with the structure; it lists its components and evaluates the effects of the ecosystem on the components. As a result, environmental engineering then remains a part of engineering, although having an awareness of ecology.

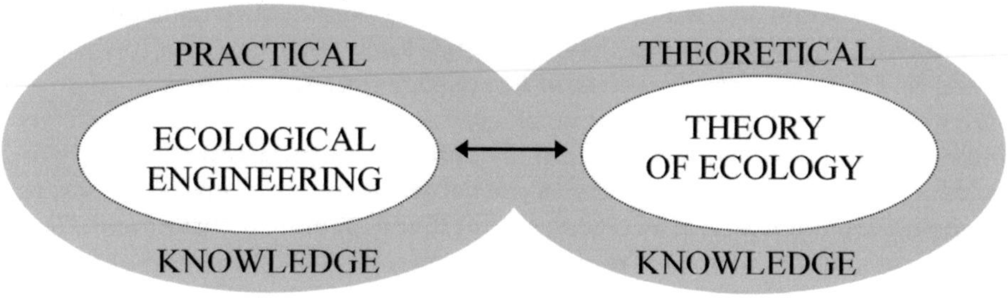

Figure 1.
Ecological engineering and its connections with the theory of ecology as well as practice and general theory.

Odum and Odum [37] maintain that environmental engineering develops the technology for connecting society to the environment. However, technology is only one part of interference with the environment. The other part is provided by the ecosystems as they organize themselves to adapt to the special conditions. Ecological engineering takes advantage of the ecosystems as they link natural resources and outputs from the economy to generate useful work.

The theory and practice, despite dialectical unity, can be distinguished by a number of specific features, among which are a degree of generalization of problems, assumptions, a subject of analysis, etc.

A goal of knowledge management is, in general, to inform and influence decision-making in the organization. Knowledge is recognized as the most important resource of the organization. In fact, maps of knowledge are helpful tools in knowledge management. They are usually created on the basis of audits [38].

The management of an organization's environmental programs in a holistic and documented manner is often called the environmental management system (EMS). In 1996, the International Organization for Standardization adopted a new international standard for EMS-ISO 14001 [39]. The actual language of the standard is that the information should be communicated to facilitate effective environmental management [40]. According to Kacsmerk [41], there are several subjects of environmental management:

- Creation of biological infrastructure, which contains all components of environment conditioning life forms on Earth
- Creation of ecological and technical infrastructure, in which all components of the natural environment dominate, as a set of conditions accompanying production and determining its proper processes
- Resources conditioning the continuity of economic processes
- Production functions, including individual components of the natural environment
- Culture-forming and civilization functions related to the impact of the natural environment on the non-economic sphere of human activity, influencing the creation of the value system of a given society

Hamdoun et al. [42] maintain that there are clear relationships between quality management, environmental management, knowledge transfer, and innovation. It can be noted that quality management has a positive effect on environmental management. Then, quality management and environmental management positively influence innovation, and what is interesting is that both quality management and environmental management positively influence knowledge transfer. It was also revealed that there is a positive effect of knowledge transfer on innovation.

A combination of "management" and "civil engineering" disciplines delivers foundations of knowledge management in construction companies. The knowledge must relate to problems connected with the nature of construction processes, whose implementation is embedded in closer and further economic environment. The management staff of construction companies must be able to use market opportunities to get involved in the implementation of construction projects in a way that ensures achieving the organization's strategic goals. They should also be able to create the operational prospects for anticipated forms and ranges of participation of the company in construction projects. Experience accumulates organizational knowledge and, along with the ability to predict economic principles, also at the global level, allows to transform construction enterprises into learning organizations. Seeing that the construction industry is increasingly competitive, and demanding improved inter-organizational relations, construction companies cannot use out-of-date business philosophies, if they want to remain in business [43]. Practical knowledge about construction projects starts with choosing the right place for buildings or nonbuilding structures. A building plot should have the right size and shape. It is also worth to check out if the location is near wetlands or floodplains and whether the plot has access to a public road. Formal issues also include a verification of the local development plan documents and other statements.

The next part of this chapter will be devoted to the relationship between ecological quality and construction processes.

5. Ecological quality of construction process ecosystem

Raising the level of environmental sensitivity leads to the implementation of environmental management principles at various levels of human activity. This applies, in particular, to the AEC industry. Construction processes consume substantial amounts of resources, (raw) materials and energy, and leave their products (buildings, roads, etc.) with many years of life, what requires special consideration of complex relationships between construction production processes and environment.

In recent times, in many countries, there is an increased interest and progress both in the theory of environmental quality management and in the practical application of new environmental management concepts in entities operating in the business environment. Practical effects are brought by the national environmental protection plans and other specific institutional measures. These effects are observed in the form of reducing pollution from various sources. An example of systemic management of environmental protection can be found in many countries. A clear pro-ecological activity, at the level of environmental quality management, is the creation of global standard regulations. The International Organization for Standardization introduced environmental standards of the ISO 14000 family. These documents, despite a lack of their mandatory character, have been widely used so far. Production systems are an essential source of ecological risk, due to the multifaceted connections with the natural environment. The progress, in which advances in technology, science, and social organization produce an improvement in entire societies, carries a number of potential environmental threats. The emerging production plants operating in the natural environment benefit from environmental goods, but unfortunately, on the other hand, are the source of emissions and waste. The outcomes of production processes are also a question mark for the environment. Relations between particular elements of production systems are presented in **Figure 2**.

The implementation of environmental management strategies is possible provided that the information about the environmental system is adequately processed. This applies to both modelling or creating mappings of elements of production systems, as well as the quality of input information, including mainly the specification of places where environmental risks are created. The methods of presenting processed information and interpreting results are also important. In particular, one can mention a way of constructing the model of the environmental impact of production, completeness of threat specifications, variability of threats, significance of the impact of threats on individual features of ecosystems, a method of estimating critical values, and data accuracy (accuracy of measurements, accuracy of readings, distortion). It must be remembered that insignificant changes slowly accumulate in tendencies, and therefore models of environmentally friendly decision-making should be dynamic. From the point of view of places of occurrence of threats that cause ecological risk, it is possible to classify environmental risk factors (externalities) of production systems, as in **Table 1**.

The ecological quality of construction production must be considered in an initial (conceptual planning) phase: e.g., by adjusting the management of the production processes to the ISO 14000 standards. These standards are a set of guidelines, which is in some descriptive documents helpful in the implementation of the so-called cleaner production. A systemic approach to managing the ecological quality of construction production is a prerequisite for obtaining positive environmental effects. A condition of effective environmental management is the systematic collection of information about a state of the environment, as well as the sources of potential hazards in production systems.

Geographic information systems (GIS) can be treated as a tool for creating a comprehensive model of these phenomena. Digital maps can be an excellent source of information for making strategic decisions in the spatial management on the regional, macro-regional, country, or international level. Such complementary data can be very useful in making decisions in environmentally managed production systems (with a significant impact on environmental protection). Modelling the ecological quality of construction production, with particular emphasis on ecological risk identification, aims to show the directions of preventive activities

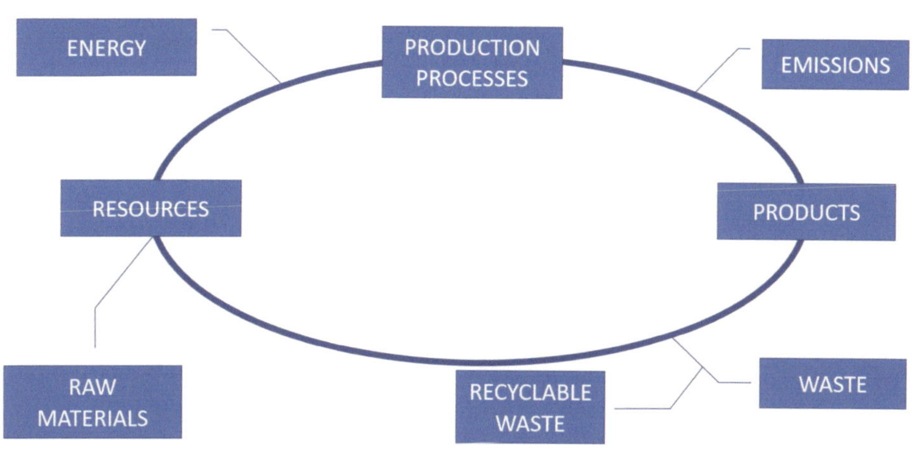

Figure 2.
Production processes ecosystem.

Type of impact factor	Place of occurrence and source of threat						
	Exterior of production system				Interior of production system		
x	Natural processes	Processes stimulated by human activity	Ecological disasters	Forces of nature	Increased consumption of natural resources	Failures of the system components	Mistakes of decision-makers
Systematic factors	Changes in the environment cumulating in trends (e.g., greenhouse effect)	Increase in emission of solid pollutants, dust, gases, radiation, noise (on input to the system)	x	Incapacity of the forces of nature to absorb waste	Depletion of resources, increase of waste	Damage to elements of technical systems	Ecological policy
Non-systematic factors	Incorrect estimation of input data to the system	Development of new techniques and technologies generating new threats	Sudden damage to technical systems in the environment (e.g., explosion at a nuclear power plant)	Anomalies of nature (e.g., floods)	Cumulative effects of resource consumption (including water, energy)	Failures of technical systems	Errors in modelling phenomena and estimating data

Table 1.
Characteristics of sources of environmental threats.

in relation to the predicted threats to ecosystems. The discovery of nature, and the place of occurrence of threats, as well as the level of risk in ecosystem modelling, is conducive to making accurate decisions in the field of environmentally friendly actions.

The implementation of environmental management principles, including ecological risk, may bring a number of effects, i.e., more efficient use of (raw) materials, and energy leading to the reduction of consumption. Improvements in manufacturing processes lead to a minimization of waste and reduction of costs and enable for the creation of new products and technologies based on environmentally friendly processes ("cleaner production" modes). Also avoiding high costs related to environmental damage (insurance premiums, costs of actions to remove damages) is another effect of intelligent environmental management. Environmental management in construction production, with particular emphasis on the identification of environmental risks, aims to show the directions of preventive actions in relation to the anticipated threats to ecosystems. In the following part of this chapter, the results of our own research on the vision of AEC as an environmentally friendly sector will be presented.

6. Eco-friendly vision of AEC: study results

6.1 Method

This research was carried out in the form of interviewing technique in which the respondent used an electronic device to answer the questions (computer-assisted personal interviewing). A pilot survey was launched on www.surveymonkey.com platform in January 2019. Thirty participants of construction processes employed by construction companies were asked to complete the questionnaire. They were supported by the researcher. A leading role of respondents is illustrated in **Figure 3**, and their experience is illustrated in **Figure 4**.

The questionnaire consisted of the two questions about a sample description, and the rest were focused on obtaining an answer consistent with the respondent's own conviction regarding the particular areas surveyed, with a degree of

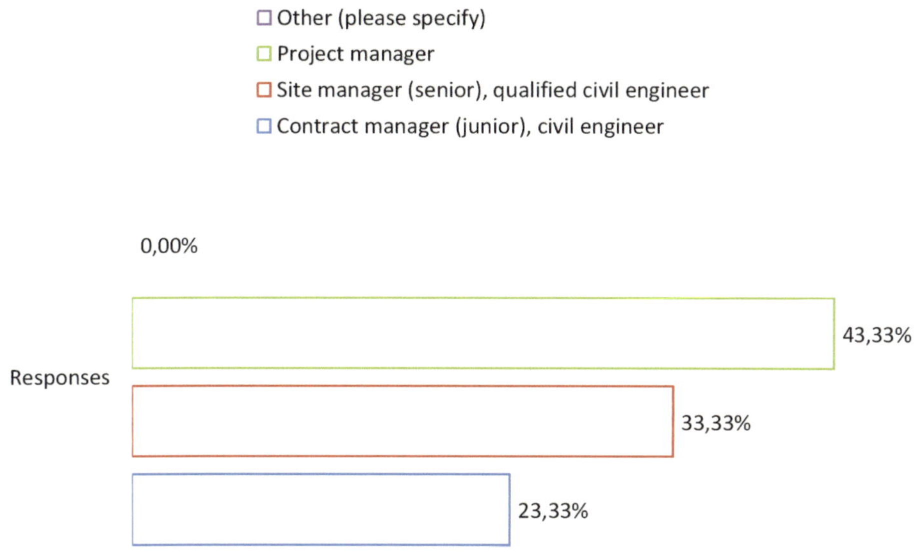

Figure 3.
Leading role of respondents.

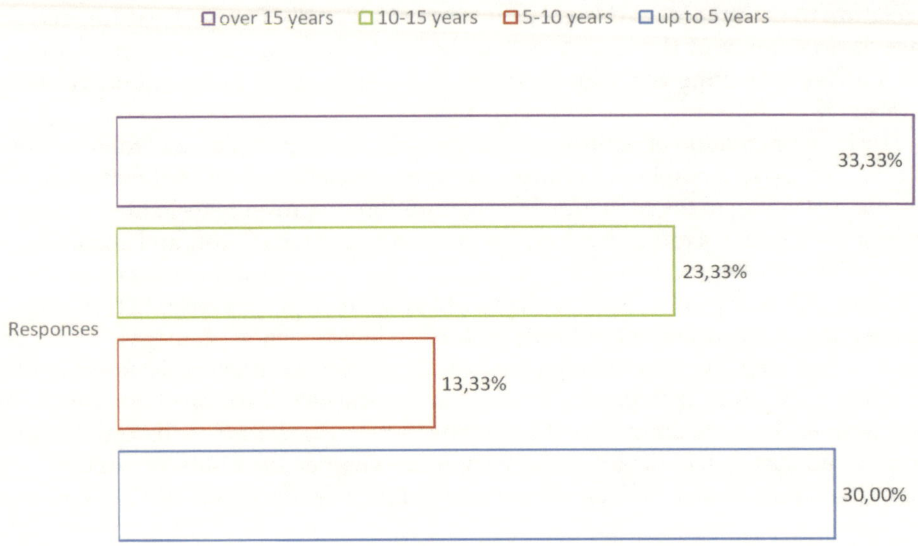

Figure 4.
Experience of respondents.

compliance on a five-point Likert scale, where "1" means "strongly disagree" and "5" "strongly agree."

6.2 Basic attributes of eco-friendly construction

The respondents commented on the basic attributes of eco-friendly construction. According to their conviction the most important are:

- Use of low-energy technologies for the construction of buildings and non-building structures (weighted average = 3,63)
- Limiting labor intensity (weighted average = 3,57)
- Ecological quality of design variants for buildings and nonbuilding structures (weighted average = 3,53)
- Use of low-cost technologies for maintenance of buildings and nonbuilding structures (weighted average = 3,53)

However, the rest are also significant (weighted average over 3,0):

- Use of renewable energy in the whole life cycle of buildings (weighted average = 3,30)
- Limiting water consumption during the entire life cycle of buildings (weighted average = 3,27)
- Use of recyclable building materials (weighted average = 3,13)
- Application of just-in-time (JIT) method in construction works (weighted average = 3,10)

In the scope of the research was also to extract knowledge about the desired individual skills expected from employees working in eco-friendly construction. The most important, according to the respondents, are:

- Ability of decision-making under risk
- Experience in project management
- Interpersonal skills
- Knowledge about building materials used in eco-friendly construction
- Knowledge about decision-making process
- Knowledge about the ecological quality of construction technology
- Knowledge about the natural environment
- Openness to innovation
- Systems thinking skills

The set of skills with their significance is shown in **Figure 5**.

According to the respondents, the most important skill demanded from employees of eco-friendly construction is the openness to innovations. However, the rest eight qualities present a similar level of significance (3,6–3,9).

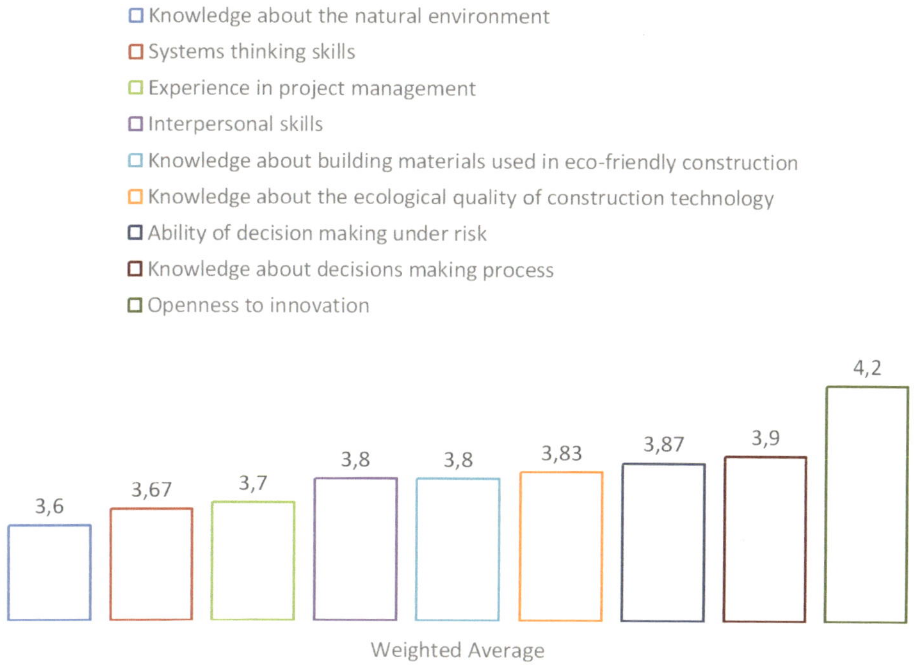

Figure 5.
Desired individual skills of employees working in eco-friendly construction.

6.3 Ecological quality factors in the opinion of participants of the construction process

6.3.1 Ecological quality in the design and construction phase

As the main factors of the ecological quality of construction processes in the design and construction phase, there are:

- Designing and accounting for water consumption
- Designing buildings according to BIM standards
- Designing low-energy houses
- Low-energy building techniques
- Organization of logistic processes according to just-in-time (JIT) criterion
- Reduction in waste of building materials
- Taking into account the idea of circular economy in the design phase
- Use of energy-saving construction machinery and equipment
- Use of recyclable building materials
- Use of reusable building materials

These results are collected in **Figure 6**.

The respondents maintain that two first phases of construction projects (design and construction) bring some difficulties in judging which factors are the most important for assessing ecological quality in projects.

6.3.2 Ecological quality in the maintenance and end-of-life phase

At the end of the research, the respondents were asked to respond to ecological quality in the maintenance and end-of-life phase. The following have been indicated as the most important:

- Complying with recommendations of building management
- Demolition of buildings with respect for circular economy requirements
- Demolition of buildings with respect for ecosystem
- Monitoring the consumption of raw materials
- Noise regulation
- Reduced energy consumption
- Reduced water consumption

- Regular building management

- Use of renewable energy

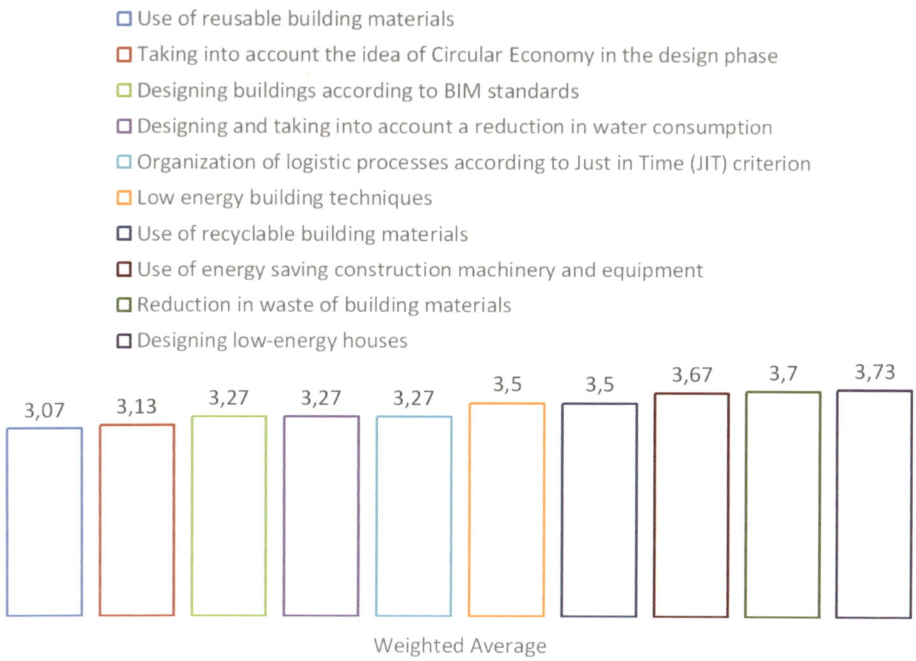

Figure 6.
Factors for assessing ecological quality in the design and construction phase.

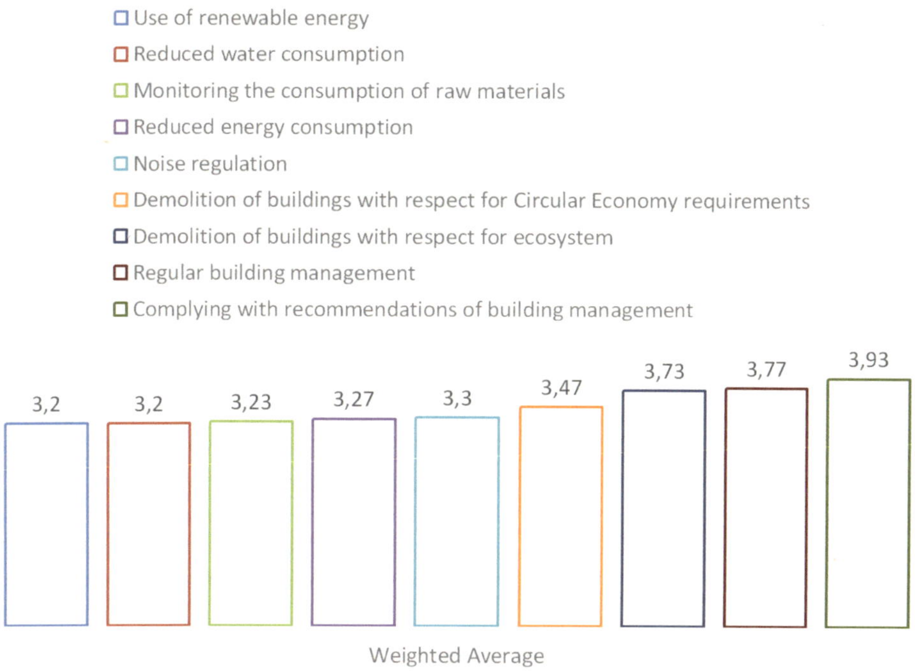

Figure 7.
Factors for assessing ecological quality in the maintenance and end-of-life phase.

These results are presented in **Figure 7**. Opinions of the respondents about the two last phases of construction projects (maintenance and end-of-life) also bring some difficulties in judging which factors are the most important to assess ecological quality in construction projects. Their weighted averages vary between 3 and 4.

One of the most important challenges for authorities and policymakers is to convince the construction market that being environmentally friendly, and becoming an eco-friendly company pays off. This requires learning innovative ecological technologies, which is to start implementing innovative processes for the cleaner production of ecological products.

Although the research is still in its embryonic stage, it gives an insight into some crucial problems connected with a hierarchy of attributes of eco-friendly construction and ecological quality factors in particular phases of construction projects.

The conducted research enabled to create the eco-friendly vision of AEC sector. In the next part of this chapter, the relationship between GPP model and CE policy in the European Union will be presented.

7. GPP model as part of European CE policy

Described before, good practice cases available online, accessible to all interested bodies responsible for public procurement were divided into 22 sections. Among them, there are eight areas directly connected with the AEC sector. These are:

- Buildings (30 cases, accessed January 2020)
- Furniture (12 cases, accessed January 2020)
- Gardening products and services (3 cases, accessed January 2020)
- Indoor lighting (4 cases, accessed January 2020)
- Office building design, construction, and management (3 cases, accessed January 2020)
- Street lighting and traffic signals (6 cases, accessed January 2020)
- Road design, construction, and maintenance (2 cases, accessed January 2020)
- Water-based heaters (2 cases, accessed January 2020)

The rest can be treated as areas indirectly connected with the AEC sector.

Thanks to the publication of information on the course of the selection process of the best offer under procurement procedures and detailed descriptions of the background of the contract, the adopted objectives, selection criteria used in tenders, obtained results, as well as the achieved environmental impacts, the European Union disseminates information on good practices that may be replicated in the future by other public institutions. The authorities may use lessons learned that are given in the reports.

More and more suggestions promoted by the European Commission are connected with circular economy. In the eight areas, mentioned before, there are five pure examples of applying CE principles to procurement procedures [21]. Two of them are coming from the Netherlands ("Circular Procurement of Furniture for the City Hall of Venlo," "Circular Procurement of Furniture for the City of

Wageningen") and one from Denmark ("Circular procurement for a sustainable learning environment," Aalborg), Sweden ("Furniture framework applying circular economy principles," Malmö), and Switzerland ("A low carbon, circular economy approach to concrete procurement," Zurich).

According to the repeating conclusions from the sustainable procurement processes, there is a need to carry out a thorough analysis of the whole process before starting the procedure. Moreover, it is necessary to collaborate closely with all stakeholders involved in the process, whereas sometimes some extra training sessions are needed to increase awareness of the business partners. However, all case studies testify to the rightness of the chosen pattern of conduct in relation to public procurement. The European Union wants to promote its own, improved over the years, economic development model among all its member states.

Nevertheless, there are different models of development seen all over the world. The key players try to adapt a need for sustainable development to local circumstances. For instance, an interesting comparison of urban planning models from Sweden and China has been published so far [44]. It seems that the European Union's model is like the Swedish one which prefers slower but more resilient development of urban areas, rather than a vertical mode, which produces fast results along with all negative consequences, including the environmental pollution and the negligence of sustainability.

The European Union, by promoting GPP, raises awareness of environmental issues among public authorities, as well as sets an example to private consumers.

The rich experience of European countries in the implementation of green public procurement, numerous examples of good practices, and the multitude of educated public clerks mean that the example of the European Union can be set as a role model for others. By promoting GPP, the European Union is developing its policy based on circular economy principles.

8. Conclusions

AEC is a sector of the economy with a significant influence on the environment. Buildings and other structures shape our surroundings and "consume" many resources throughout their life cycle. Contractors have to be sensitive to environmental issues.

In the chapter, based on considerations taken from the literature review as well as direct interviews with experts of the construction sector, it was revealed that knowledge management system in every construction company should cover also, and maybe primarily, the environmental knowledge. In order to indicate significant contents of such knowledge, a survey was conducted among construction engineering experts. The respondents pointed out the subjective role of companies and described it as crucial, indicating a number of individual skills required in eco-friendly construction. The study allowed to discover the buildings' life cycle approach to the creation of environmental knowledge of construction companies.

The chapter identifies the circular economy as an element of the strategic policy of the European Union. Treated as an effective mechanism for sustainable development, CE has become a pillar of GPP.

Despite the nonobligatory nature of the rules related to GPP, the European Union focuses on educating decision-makers, directly public and indirectly private ones. It is worth noting that the GPP model includes not only CE but a number of other solutions supporting sustainable development. The European Union policy results from the need to respond to the deteriorating condition of the natural environment. The growing environmental threats from industry and services require an immediate response. However, changes in improving production conditions take time.

On the other hand, there are often numerous restrictions affecting the risk of such activities. The most serious threats include the low adaptability of other players, limited knowledge of sustainable development, GPP and CE, as well as reluctance to change. It seems that one of the most serious risk factors—apart from those mentioned earlier—is the routine of public authorities and the lack of willingness to go beyond the usual framework of existing legal procedures related to public procurement.

Sometimes safety, provided by well-established patterns of conduct, can be illusory. It is worth taking a risk and turning towards GPP, which give the opportunity to achieve even better results than before.

To use the full potential of GPP, along with many environmentally friendly mechanisms (including CE), one should use the model promoted in the European Union and presented in this chapter. Some decision-makers can share their experience with others. In addition, a crucial remark is that the cooperation of all participants of investment and construction projects and all players from the AEC sector is necessary.

It is worth remembering that contemporary economic activity has an impact on these and future generations. Sometimes it is worth considering how we can stop the processes that have a negative impact on the natural environment. Maybe it is worth thinking about GPP, maybe CE is not an odd idea, especially when the temperature outside is positive, although it is usually frost and snow.

Conflict of interest

The authors declare no conflict of interest.

Thanks

Many thanks to professor Pedro Núñez-Cacho Utrilla from the University of Jaén (Spain) who inspired me to treat the construction sector as a subject of the circular economy thinking.

Moreover, I would like to thank IntechOpen editorial office who helped in handling the article preparation process.

Author details

Jarosław Górecki
UTP University of Science and Technology, Bydgoszcz, Poland

*Address all correspondence to: gorecki@utp.edu.pl

References

[1] Wyrobisz A. Les Metiers du Batiment en Petite-Pologne au XIV et au XV siecle | Budownictwo Murowane w Małopolsce w XIV i XV Wieku. Warsaw: Ossoliński National Institute; 1963. p. 170

[2] Geyer R, Jambeck JR, Law KL. Production, use, and fate of all plastics ever made. Science Advances. 2017;**3**(7):1-5

[3] Krivoshapko SN, Shambina SL, Hyeng CAB. Thin-walled composite and plastic shells for civil and industrial buildings and erections. In: Materials Science Forum. 2017. pp. 45-51

[4] Worm B, Lotze HK, Jubinville I, Wilcox C, Jambeck J. Plastic as a persistent marine pollutant. Annual Review of Environment and Resources. 2017;42:1-26

[5] Ostle C, Thompson RC, Broughton D, Gregory L, Wootton M, Johns DG. The rise in ocean plastics evidenced from a 60-year time series. Nature Communications. 2019;**10**(1):1-6

[6] Alimba CG, Faggio C. Microplastics in the marine environment: Current trends in environmental pollution and mechanisms of toxicological profile. Environmental Toxicology and Pharmacology. 2019;**68**:61-74

[7] European Commission. Public procurement for a better environment [Internet]. Belgium; 2008. Available from: https://eur-lex.europa.eu/legal-content/EN/TXT/?uri=CELEX:52008DC0400

[8] Ellen MacArthur Foundation. What is a circular economy? [Internet]. Available from: https://www.ellenmacarthurfoundation.org/circular-economy/concept

[9] Núñez-Cacho P, Górecki J, Molina-Moreno V, Corpas-Iglesias FA. New measures of circular economy thinking in construction companies. Journal of EU Research in Business. 2018;**2018**:1-16

[10] Khasreen M, Banfill PF, Menzies G. Life-cycle assessment and the environmental impact of buildings: A review. Sustainability. 2009;**1**(3):674-701

[11] Nuñez-Cacho P, Górecki J, Molina-Moreno V, Corpas-Iglesias F. What gets measured, gets done: Development of a circular economy measurement scale for building industry. Sustainability. 2018;**10**(7):2340

[12] Górecki J, Núñez-Cacho P, Corpas-Iglesias FA, Molina V. How to convince players in construction market? Strategies for effective implementation of circular economy in construction sector. Cogent Engineering. 2019;**6**(1)

[13] Hossain MU, Ng ST. Critical consideration of buildings' environmental impact assessment towards adoption of circular economy: An analytical review. Journal of Cleaner Production. 2018;**205**:763-780

[14] Górecki J. Simulation-based positioning of circular economy manager's skills in construction projects. Symmetry. 2020;**12**(1):1-25. Available from: https://www.mdpi.com/2073-8994/12/1/50

[15] de Abreu MCS, Ceglia D. On the implementation of a circular economy: The role of institutional capacity-building through industrial symbiosis. Resources, Conservation & Recycling. 2018;**138**(July):99-109

[16] Whicher A, Harris C, Beverley K, Swiatek P. Design for circular economy: Developing an action plan for Scotland.

Journal of Cleaner Production. 2018;**172**(December 2015):3237-3248

[17] Pawlyn M. Biomimicry in Architecture. Newcastle upon Tyne: RIBA Publishing; 2016

[18] Nasir MHA, Genovese A, Acquaye AA, Koh SCL, Yamoah F. Comparing linear and circular supply chains: A case study from the construction industry. International Journal of Production Economics. 2017;**183**:443-457

[19] Silvestre JD, De Brito J, Pinheiro MD. Environmental impacts and benefits of the end-of-life of building materials—Calculation rules, results and contribution to a "cradle to cradle" life cycle. Journal of Cleaner Production. 2014;**66**:37-45

[20] Jimenez-Rivero A, Garcia-Navarro J. Best practices for the management of end-of-life gypsum in a circular economy. Journal of Cleaner Production. 2017;**167**:1335-1344

[21] European Commission. Green Public Procurement [Internet]. Available from: https://ec.europa.eu/environment/gpp/case_group_en.htm

[22] European Commission. Buying Green! A Handbook on Green Public Procurement. 2016

[23] European Commission. Directive 2014/24/EU of the European Parliament and of the Council. Official Journal of the European Union; 2014

[24] European Commission. EU GPP Criteria for Sanitary Tapware [Internet]. 2013. Available from: https://ec.europa.eu/environment/gpp/pdf/criteria/sanitary/EN.pdf

[25] European Commission. EU GPP Criteria for Flushing Toilets and Urinals [Internet]. 2013. Available from: https://ec.europa.eu/environment/gpp/pdf/criteria/toilets/criteria_Toilets_en.pdf

[26] European Commission. Green Public Procurement Criteria for Waste Water Infrastructure [Internet]. 2013. Available from: https://ec.europa.eu/environment/gpp/pdf/waste_water_criteria.pdf

[27] European Commission. EU GPP Criteria for Water-based Heaters [Internet]. 2014. Available from: https://ec.europa.eu/environment/gpp/pdf/criteria/water_based/heaters_en.pdf

[28] European Commission. EU Green Public Procurement Criteria for Road Design, Construction and Maintenance [Internet]. 2016. Available from: https://ec.europa.eu/environment/gpp/pdf/GPPcriteriaRoads(2016)203.pdf

[29] European Commission. EU GPP Criteria for Office Building Design, Construction and Management [Internet]. 2016. Available from: https://ec.europa.eu/environment/gpp/pdf/swd_2016_180.pdf

[30] European Commission. EU green public procurement criteria for paints, varnishes and road marking [Internet]. 2017. Available from: https://ec.europa.eu/environment/gpp/pdf/criteria_for_paints_varnishes_and_road_marking.pdf

[31] European Commission. EU green public procurement criteria for road lighting and traffic signals [Internet]. 2018. Available from: https://ec.europa.eu/environment/gpp/pdf/toolkit/181210_EU_GPP_criteria_road_lighting.pdf

[32] European Commission. EU green public procurement criteria for road transport [Internet]. 2019. Available from: https://ec.europa.eu/transparency/regdoc/rep/10102/2019/EN/SWD-2019-2-F1-EN-MAIN-PART-1.PDF

[33] European Commission. EU green public procurement criteria for public space maintenance [Internet]. 2019. Available from: https://ec.europa.eu/environment/gpp/pdf/191113_EUGPPcriteriaforpublicspacemaintenance_SWD(404)2019final.pdf

[34] Mitsch WJ, Jørgensen SE. Ecological engineering: A field whose time has come. Ecological Engineering. 2003;**20**(5):363-377

[35] Gattie DK, Smith MC, Tollner EW, McCutcheon SC. The emergence of ecological engineering as a discipline. Ecological Engineering. 2003;**20**(5):409-420

[36] Allen TF, Giampietro M, Little A. Distinguishing ecological engineering from environmental engineering. Ecological Engineering. 2003;**20**(5):389-407

[37] Odum HT, Odum B. Concepts and methods of ecological engineering. Ecological Engineering. 2003;**20**(5):339-361

[38] Mearns MA, du Toit ASA. Knowledge audit: Tools of the trade transmitted to tools for tradition. International Journal of Information Management. 2008;**28**(3):161-167

[39] Melnyk SA, Sroufe RP, Calantone R. Assessing the impact of environmental management systems on corporate and environmental performance. Journal of Operations Management. 2003;**21**(3):329-351

[40] Cheremisinoff NP, Bendavid-Val A. Green profits. In: The Manager's Handbook for ISO 14001 and Pollution Prevention. 1st ed. Butterworth-Heinemann; 2001. p. 356

[41] Kaczmarek B. Policy and strategy for ecological development of the enterprise—A sketch of the problem. Civil and Environmental Engineering. 2011;**2**:507-510

[42] Hamdoun M, Chiappetta Jabbour CJ, Ben Othman H. Knowledge transfer and organizational innovation: Impacts of quality and environmental management. Journal of Cleaner Production. 2018;**193**:759-770

[43] Holt GD, Love PED, Li H. The learning organisation: Toward a paradigm for mutually beneficial strategic construction alliances. International Journal of Project Management. 2000;**18**(6):415-421

[44] Wennersten R. Development of new sustainable urban areas: Horizontal or vertical planning systems for resource efficient cities. In: Ergen Y, editor. An Overview of Urban and Regional Planning. London: IntechOpen; 2018. pp. 103-120

Chapter 7

Impact of Zero Energy Building: Sustainability Perspective

Wesam Salah Alaloul and Muhammad Ali Musarat

Abstract

In an era with major developments in the energy sector, along with many benefits of energy consumption, it is also showing adverse effects on the end-users and the environment due to emission of various harmful gases mainly carbon dioxide (CO_2). To deal with these issues, the zero energy building emerges to bring constructive developments through the construction industry. The concept of zero energy building is to develop a structural building which can generate its own required energy and have zero negative effects. The energy will be enough to fulfill all the requirements of the building operations and can save natural quarries. By increasing the numbers of zero energy buildings, major reforms can be brought in the construction industry and thus stabilizing the economy and the climate.

Keywords: energy consumption, harmful gases, CO_2, zero energy building, economy, climate

1. Introduction

The energy sector is going through numerous challenges which will get worse with time. Various concerns have been reported related to the environment, economic instability, and energy security, mainly due to the present behaviour of the energy sector and carbon emissions [1]. In today business world, energy becomes a major source of economic growth. A smooth service for residential and commercial buildings involves extensive energy consumption. In this sector, energy consumption is escalating progressively which results in the emission of greenhouse gases. That is why saving energy with a suitable alternative in providing a better lifestyle gets essential. In this regard, the zero energy building is a very useful solution [2]. By integrating energy efficiency in buildings, sustainable development can be brought into the building sector. For achieving the goal of zero energy building, the design should be such that it can optimize maximum outdoor weather conditions [3].

Major reforms are brought into the construction industry for the betterment of the end-users in which zero energy building is one of them. Still, many are unaware of the concept of zero energy building as it is newly emerged area. Zero energy building is a structural element embraced for residential and commercial purposes which fulfilled the energy requirements by their own energy generation. It is very helpful in meeting the comfort requirements of the end-users, fulfilling the growing energy demand and beneficial to reduce the threat to climate changes due to global warming. Moreover, by adapting zero energy building, natural quarries can be saved from getting vanished. Therefore, this chapter discusses why there is a need for zero energy building and how it can bring reforms to real-world problems.

2. Energy demand

In the current decade, the demand of energy boosts up worldwide to 2.3% compared to the year 2018, making exceptional performance led by a vigorous global economy and high demand of heating and cooling systems in various regions. The highest consumption in the energy sector was made by natural gas due to high demand, posting 45% of the rise. With time, the demand for all the fuels getting increased where fossil fuels have a growth rate of 70%. A double pace was observed in solar energy generations which got increased by 31%, still not enough to meet the higher electricity demand. The increasing energy demand results in high carbon dioxide (CO_2) emissions escalated by 1.7% which is 33 Gigatons in the year 2018. Electricity demand increases by 4% in the year 2018 and remains to spot as the fuel of the future. In the total energy consumption, electricity contributes up to 20% [4]. **Figure 1** shows the historic and predicted data of world energy consumption.

Gradually, the energy demand getting increase, due to the global population increase, and the resources get lesser, requires an approach to overcome these phenomena. As a result, zero energy building is the most appropriate to accommodate the increased population and minimize the adverse effects of the energy shortage. In zero energy building, the energy loads are reduced up to a greater extent so that the renewable energy can meet the remaining requirements of the building, thus fulfilling the demand of end-user.

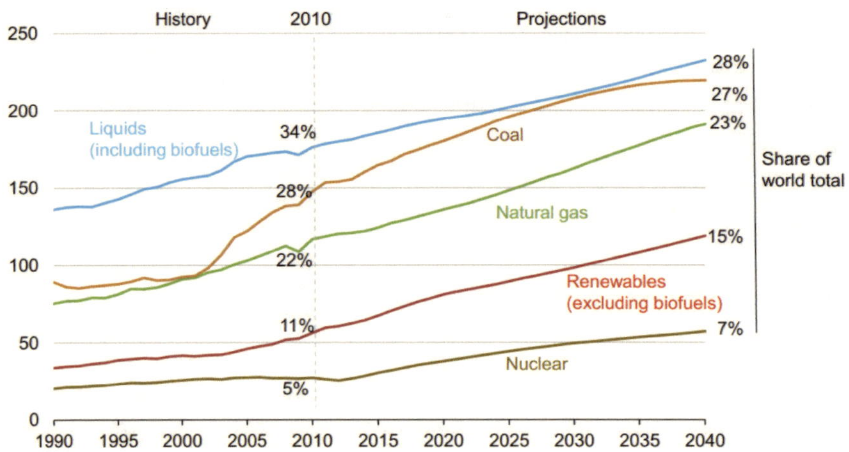

Figure 1.
World energy consumption, 1990–2040 [5].

3. Threat to climate

It is of the high interest for construction stakeholders, end-users, and the government that the construction and commissioning process should be energy-efficient and eco-efficient. Due to high energy consumption, the impact on the environment is greater, however, has been overlooked for years [6]. Sun is the energy source from ages for both humans and the other species where greenhouse gases kept the climate mild for living. But with time, these gases are threatening the living, and severe changes have occurred in the atmosphere. **Figure 2** shows the intensive amount of harmful gases erupted from a single industrial unit.

CO_2 is one of the most harmful and widespread gases in the greenhouse. The highest level of CO_2 is reported in the atmosphere mainly due to the burning of fossil fuels by humans. These gases absorb the solar energy keeping the heat over the earth surroundings instead of allowing it to evade. This phenomenon is known as the greenhouse effect. Climate change not only refers to the rise in the temperature but also to severe weather conditions which directly impacts the population and has other serious consequences as well. CO_2 itself responsible for three-quarters of the gas emissions as it remains in the atmosphere for a thousand years [8]. The greenhouse effect is explained in **Figure 3**.

The impact of greenhouse gas emissions is directly on the country's economy, civilization, and the atmosphere. The emission of CO_2 to the atmosphere breaks all the record with 410 parts per million thresholds in the world. The cause of this constant expansion is mainly due to human actions that are undermining the climate [10]. It is due to industrial revolution, which is adding CO_2 abruptly to the climate. As a result, around 1°C temperature has been increased, and the sea levels are getting higher. The impact of these changes can be seen worldwide. Beside these consequences, high heat waves, heavy rainfall, and the large wildlife distinction are

Figure 2.
Harmful gases emission from industrial unit [7].

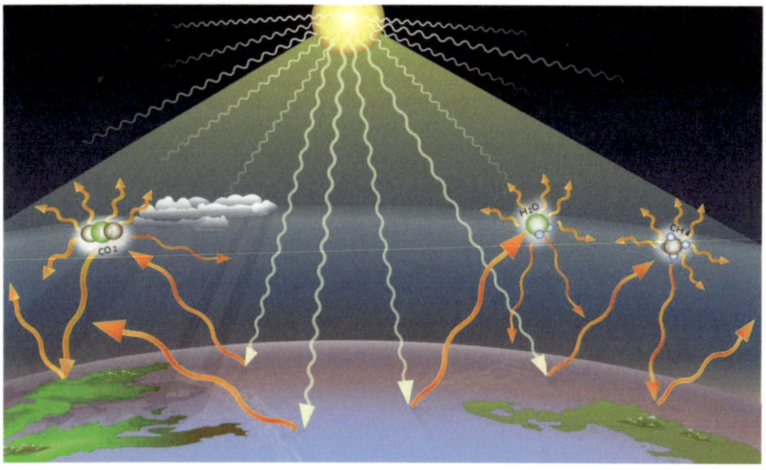

Figure 3.
Green house effect [9].

also disrupting the climate, all due to increasing temperature [11]. To portray a true picture of these effects is the current fire issue in the Australia, which occurs due to climate changes and results in assassination the life of many species, humans, and a big burden to the economy.

4. Concept of zero energy building

Reviewing the method of construction in the construction industry, the innovation is lesser compared to other industries. Previously decisions were made to get a construction performed on the lowest initial cost without giving any attention to the limitation of the resources, especially during the operation stage. Progressively, the advancement came in the industry, and the focus was to improve the properties of the available materials for better utilization and cost-efficient. The change in the philosophy of construction industry materializes to build and construct living societies with improved health and environmental conditions [12]. The construction industry has been criticized for being the major contributor to the carbon emission and global warming. Around 10% of worldwide energy is consumed while manufacturing building materials and also generates 40% of the solid waste [13]. The rising energy demand and environmental concerns lead to sustainability by providing the living facilities which minimize the harmful effects and can easily be implemented [14].

The concept of achieving zero net energy consumption and zero carbon emissions of a building is known as zero energy building, also called as a net-zero energy building. Zero energy building generates its energy resources without relying on energy grid supply. The net-zero design principle provides the ease to the building users even in the extreme conditions, the more extreme exposure to the elements the higher energy requires for the comfort [15]. This principle is getting significant attention as developed renewable energy eliminates greenhouse gas emissions [16, 17].

The growth in zero energy building mainly occurs due to advancement in construction technologies and due to the input of academic researchers by collecting and analyzing the accurate energy performance information. Though zero energy building is still not common yet gaining value in developed countries. In the current

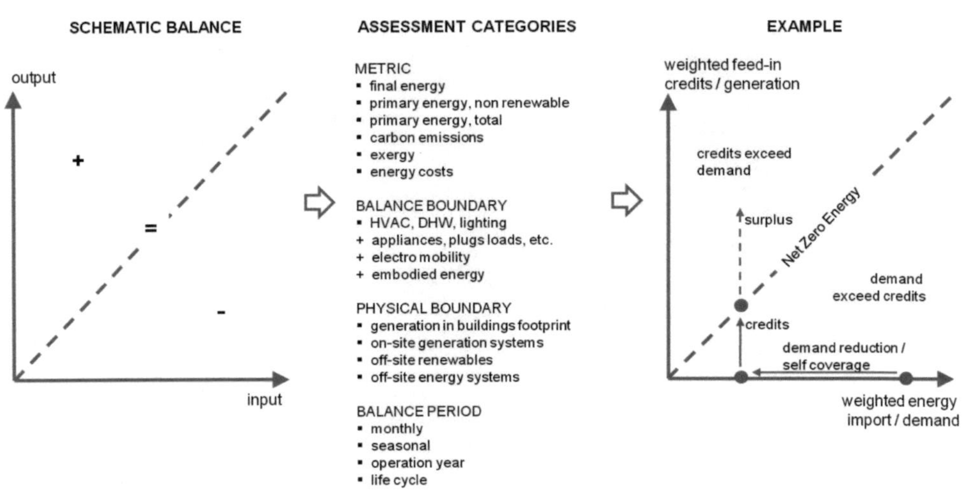

Figure 4.
Zero energy building balance concept [18].

era, computer models can detect how efficient is the engineering design decisions. With the concept of zero energy building, carbon emissions and fossil fuels dependency can be reduced [15]. The first consideration for zero energy building was made by the US Department of Energy; whereas, the European Union was the first region to mandate zero energy building use [19]. The zero energy building balance concept is described in **Figure 4**.

In real-world industrial estate, the zero energy building is the next revolution and contest. The construction industry is under pressure to provide efficient, cost-effective, and low energy consumption buildings in lesser time. In energy efficient approaches, zero energy building plays a vital role. A total of 30–40% of energy is utilized by the building sector, and reform in this area is the key step towards the future of sustainability. But this reform cannot be possible without the support of the stakeholders and should have familiarization about such kind of projects [11].

5. Construction of zero energy building

The aim is to construct a zero energy building that utilizes natural resources, lessen the waste, and fully optimize the producible energy. The construction team usually consists of engineers, architects, developers, owners, builders, and the occupants. The approach to constructing a zero energy building is to consider the building as one energy system in which every part should be energy efficient. Only those construction materials, developing systems and assemblies are taken into the account which decreases the energy use and utilizes all the built renewable energy. The building is furnished with the robust thermal envelope, providing a continuous air and moisture barrier, enhancing the effectiveness, and providing a relaxing indoor atmosphere. Site orientation is a critical factor for zero energy building as the moto is to take full benefit of the energy produced by the sun. Preferably, the orientation should be north-south in the Northern

Figure 5.
Zero energy building [20].

Hemisphere, as it lowers the direct sunlight in the summer which reduce the cooling demand and higher the sunlight in the winter to reduce heating demand. The windows should be able to utilize the maximum natural light, control the heat variations, and automatically get darken when hit by the sunlight. Moreover, southern facing windows can prevent the heat in summer and warming up in the winter using shades and louvers. The roof of zero energy building holds the building cool by preventing the heat gained by the solar panels. Thicker and light colour materials are good to keep the roof cool as they oppose the sunlight and improve the indoor conditions. As zero energy building is air tightened, a proper energy recovery ventilation system is provided which keeps the air fresh and reduces the energy losses. It is recommended to keep the connection of zero energy building with the conventional energy source as well just in case of renewable energy cannot fulfill the requirements of the end-user. Also, if the energy generated is in surplus, it can be transferred to the grid so the energy inside the zero energy building should be steady [21]. **Figure 5** shows a model of a zero energy building.

6. Pros and cons of zero energy building

There are many advantages of the zero energy building, yet everything comes with a downside. The pros and cons of the zero energy building are discussed below:

6.1 Pros

1. Due to improved energy efficiency, the cost to the end-user get reduced.

2. The comfort of zero energy building is more as compared to the conventional building due to the uniform inside atmosphere.

3. No impact of the external energy crisis to the end-user.

4. Reduction in monthly living expenses.

5. Environmentally friendly and reduce the carbon emission.

6. Higher resale value compared to conventional building.

6.2 Cons

1. Less availability of experienced designers to build zero energy building.

2. The initial cost of zero energy building is higher compared to the conventional building.

3. Not suitable in the region with cold temperature due to less exposure to the climate.

4. Limit future ability to respond to global warming due to specific temperature design.

7. Ecological restoration by zero energy building

Eventually, every industry contributes to the emission of CO_2, and construction industry is not exempted. The best way to reduce CO_2 emissions is to avoid the burning of fossil fuels. Just avoiding the burning is not enough as energy is required in all the sectors to perform day to day operation. Here, the importance of the zero energy building emerges as it is the most suitable way to avoid the emission of CO_2 and also fulfill the requirements of the end-users by generating their energy. Zero energy building is also known as the zero-carbon building as the emissions of carbon from fossil fuels get balanced with the amount of produced renewable energy [22].

Under crucial circumstances of climate change, the construction industry requires to construct high performance buildings where zero energy buildings are the robust solution as it provides healthy and energy efficient buildings which generates their own energy for usage. With the help of zero energy building, country's economy will also boost [11]. Climate change and shortage of natural resources is a global issue where adapting zero energy buildings can be restored and lemmatize the hazards.

8. Zero energy building as cost efficient

Cost efficiency of the zero energy building implies the energy cost that is utilized by the building, which is the main concern of most of the end-users. Infrastructural components and high demand costing by utility suppliers often included in energy cost. That is why cost does not only portray the energy consumed vs. energy generated by the building [18]. The main hurdle in endorsing the zero energy building is the initial construction cost which is paid by the investors. Direct and indirect costs are involved in the construction of the zero energy building. Direct cost includes materials cost, labour cost, machinery cost, and other costs which are directly related to construction activities. Indirect cost includes documentation fees, design cost, commission, and other official fees; whereas, the post-construction cost includes operational costs of building utilized in energy development [23–25].

Usually, people compare the initial construction cost of the zero energy building with the conventional building, which is higher for zero energy building, but the running expense is much lesser as compared to the conventional building. In zero energy building, all the energy demand is fulfilled by the building itself which is more cost-efficient compared to energy generated by the government, as that includes taxes and other hidden costs. It can be concluded that in the long run, zero energy buildings are way more advanced and cost-efficient, compared to conventional buildings.

9. Impact of zero energy building on economy

Zero energy building implies a significant impact on a country's economy. Every country is struggling to produce a generous amount of energy to meet the requirement of the end-users. But due to limited resources, it is getting difficult and burdened the economy as well. Zero energy building comes as a solution not only to fulfill the energy demand but also stabilize the country's economy.

There could be one perception that instead of supporting the economy, zero energy building will leave a negative impact on the economy as people will not pay taxes for energy usage. This perception arises because taxes are the main source of income for government and country's development. The perception can be refused as the government is utilizing more money for energy generation compared to the money getting in return. Zero energy building mainly operates with solar panels to generate energy. Due to this fact, the demand of the solar panels increases, so as its industry and the country's economy.

10. Social impact of zero energy building

Uncertainty in foreseeing the energy use in the building sector is due to occupant behaviour which is the most critical factor. Variation in energy usage in buildings has been observed even with the same climate conditions. The comfort level of every human being varies which directly affect the building operations and also increases the energy demand [26].

Humans are spending 90% of their lives in indoors premises for various works or living. Hence, to maintain a healthy lifestyle, safe and comfortable environment in buildings is significant. To provide the comfort and enhancing the condition of a building, almost 40% of the world's energy is been consumed which results in one-third greenhouse gas emissions mainly associated with the building sector [27, 28].

Besides environmental and economic benefits, zero energy building shows a positive impact on the society as well. Most of the benefits are related to the health of the end-users involved in working or living in the zero energy building. People associated with zero energy building tend to have an increase in brain functioning, getting better sleep at night, and due to low concentration of CO_2 and other pollutants, the overall performance also gets an increase. Zero energy building not only focuses on the environmental perspective but also aims to provide a comfortable and healthy lifestyle [29].

11. Life cycle cost analysis of zero energy building

In long run, the zero energy building is more cost-efficient compared to conventional building. A life cycle cost analysis (LCC) was performed for 20 years using present worth analysis between conventional building and zero energy building [30]. The comparison is discussed in **Table 1**.

From **Table 1**, it can be observed that although the initial cost of zero energy building is higher compared to conventional building, LCC shows that zero energy building is much more cost-efficient and economic.

12. Conclusion

High energy consumption is a threat to climate, and the changes occurring are adverse for the life on earth as it is causing global warming. Not only the plants and animals but humans are also getting affected. With time, even the sources are getting shorter for mankind and one day will vanish. To deal with this issue, alternative solutions are required which fulfill the energy demand and have no impact on the environment. In this scenario, the zero energy building emerges as the best available solution to control both the major issues. In the long run, zero energy buildings are more cost-efficient and contribute to the country's economy as well. Though understanding of zero energy building is still lesser to many but will get a boost as it is in favour of everyone.

Activity	Conventional building	Zero energy building
Construction cost (USD)		
Construction cost for building	30143.50	39339.04
Cost for home appliances	203,320	2270.04
Cost for gas arrangements	42.06 (natural gas)	42.06 (biogas)
Total initial cost	**33036.15**	**41651.15**
Operation, maintenance and replacement costs (USD)		
Present value of water charge for 20 years	5.54	5.54
Present value of electric charge for 20 years	6820	—
Present value of fuel cost for 20 years	1642.47	96.04
Present value of home appliances for 20 years	3949.64	3652.55
Present value maintenance of building	3068.19	1840.86
Present value of total operating maintenance and replacement cost for conventional building	15485.84	5594.98
LCC for conventional building	**48521.99**	**47246.13**

Table 1.
Comparison of conventional building and zero energy building.

Acknowledgements

The authors would like to thank Universiti Teknologi PETRONAS (UTP) for the support provided for this research.

Conflict of interest

The authors declare no conflict of interest.

Author details

Wesam Salah Alaloul* and Muhammad Ali Musarat
Department of Civil and Environmental Engineering, Universiti Teknologi PETRONAS, Tronoh, Perak, Malaysia

*Address all correspondence to: wesam.alaloul@utp.edu.my

References

[1] Waide MP, Gerundino MD. International standards to develop and promote energy efficiency and renewable energy sources. In: Prepared for the G8 Plan of Action. IEA Information Paper. Retrieved 28 March 2007. p. 2011

[2] Delavar H, Sahebi H. A sustainable mathematical model for design of net zero energy buildings. Heliyon. 2020;**6**(1):e03190

[3] Feng W et al. A review of net zero energy buildings in hot and humid climates: Experience learned from 34 case study buildings. Renewable and Sustainable Energy Reviews. 2019;**114**:109303

[4] Iea. Global energy demand rose by 2.3% in 2018, its fastest pace in the last decade. 2019. Available from: https://www.iea.org/news/global-energy-demand-rose-by-23-in-2018-its-fastest-pace-in-the-last-decade

[5] W. Commons. World energy consumption, 1990-2040, EIA Energy Outlook. 2013. Available from: https://commons.wikimedia.org/wiki/File:World_energy_consumption,_1990-2040,_EIA_Energy_Outlook_2013.png

[6] Li W. Efficiency of Manufacturing Processes. Germany: Springer; 2015

[7] Piqsels. Available from: https://www.piqsels.com/en/search?q=pollution&page=4

[8] Nunez C. Carbon dioxide levels are at a record high. Here's what you need to know. Available from: https://www.nationalgeographic.com/environment/global-warming/greenhouse-gases/

[9] Wikipedia. Greenhouse effect. Available from: https://en.wikipedia.org/wiki/Greenhouse_effect

[10] Berry P, Sánchez-Arcilla Conejo A, Betts R, Harrison PA. High-End Climate Change in Europe: Impacts, Vulnerability and Adaptation. Sofia, Bulgaria: Pensoft Publishers; 2017

[11] Attia S. Net Zero Energy Buildings (NZEB): Concepts, Frameworks and Roadmap for Project Analysis and Implementation. UK: Butterworth-Heinemann; 2018

[12] Marjaba G, Chidiac S. Sustainability and resiliency metrics for buildings–critical review. Building and Environment. 2016;**101**:116-125

[13] Wong JKW, Zhou J. Enhancing environmental sustainability over building life cycles through green BIM: A review. Automation in Construction. 2015;**57**:156-165

[14] Azhar S, Carlton WA, Olsen D, Ahmad I. Building information modeling for sustainable design and LEED® rating analysis. Automation in Construction. 2011;**20**(2):217-224

[15] E. A. ECOLOGY. Zero-Energy Buildings. Available from: http://environment-ecology.com/energy-and-architecture/152-zero-energy-buildings.html

[16] Baden S, Fairey P, Waide P, de T'serclaes P, Laustsen J. Proceedings of Hurdling Financial Barriers to Low Energy Buildings: Experiences from the USA and Europe on Financial Incentives and Monetizing Building Energy Savings in Private Investment Decisions. Washington, DC: American Council for an Energy Efficient Economy; 2006. pp. 5-8

[17] U. D. O. Energy. Annual Energy Review. 2008. Available from: https://www.eia.gov/totalenergy/data/annual/index.php

[18] W. Commons. Net ZEB balance concept. Available from: https://commons.wikimedia.org/wiki/File:Net_ZEB_balance_concept.png

[19] Iyer-Raniga U. Zero energy in the built environment: A holistic understanding. Applied Sciences. 2019;**9**(16):3375

[20] Flicker. Available from: https://www.flickr.com/photos/44221799@N08/4647853823

[21] Bautex. 10 Tips Architects and Builders Use To Build A Net-Zero Energy Office Building. Available from: https://www.bautexsystems.com/blog/net-zero-energy-office-building

[22] CIC. Overview of ZCB. 2017. Available from: http://www.cic.hk/eng/main/zcb/ZCB_experience/Overview_of_ZCB/

[23] Khoshbakht M, Gou Z, Dupre K. Cost-benefit prediction of green buildings: SWOT analysis of research methods and recent applications. Procedia Engineering. 2017;**180**:167-178

[24] Yudelson J. The Green Building Revolution. Washington, DC: Island Press; 2010

[25] Hu M. Does zero energy building cost more?–an empirical comparison of the construction costs for zero energy education building in United States. Sustainable Cities and Society. 2019;**45**:324-334

[26] Barthelmes VM, Becchio C, Corgnati SP. Occupant behavior lifestyles in a residential nearly zero energy building: Effect on energy use and thermal comfort. Science and Technology for the Built Environment. 2016;**22**(7):960-975

[27] U. C. P. S. Commission. The Inside Story: A Guide to Indoor Air Quality. Washington, DC: US Environmental Protection Agency; 1993

[28] Sbci U. Buildings and climate change: Summary for decision-makers. In: United Nations Environmental Programme, Sustainable Buildings and Climate Initiative, Paris. 2009. pp. 1-62

[29] W. G. B. Council. The benefits of Green Buildings. Available from: https://www.worldgbc.org/benefits-green-buildings

[30] Anju MS. Comparison of cost and energy efficiencies of zero energy residential building and conventional building. International Journal of Engineering and Science Invention (IJESI). 2017;**6**(7):64-71